AN INTIMATE LOOK AT THE

Night Sky

"It is breathtaking

simply

to be here"

—Rainer Maria Rilke

An Intimate Look at the *Night Sky*

CHET RAYMO

WALKER & COMPANY
NEW YORK

First published in the United States of America in 2001 by
Walker Publishing Company, Inc.

Published simultaneously in Canada by Fitzhenry and Whiteside,
Markham, Ontario L3R 4T8

The poems on pages 50 and 102–3 are from *The Poems of Gerard Manley Hopkins*, edited by W. H. Gardner and N. H. MacKenzie, Oxford University Press, Inc., New York, revised 1967.

All star maps created by Dan Raymo. Maps from Starry Night Pro astronomy software were adapted using Corel Draw 9.0 and Corel Draw Clipart Collection, which are protected by the copyright laws of the United States, Canada, and elsewhere. Used under license.

Library of Congress Cataloging-in-Publication Data
available upon request
ISBN 0-8027-1369-6

Book design by Mspaceny/Maura Fadden Rosenthal

Printed in the United States of America
10 9 8 7 6 5 4 3 2 1

Contents

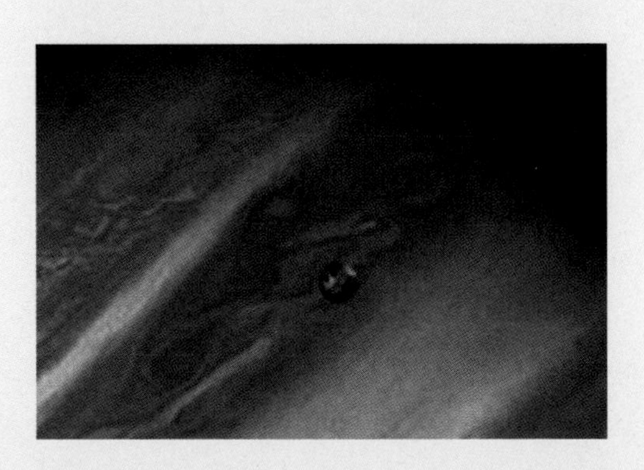

III. Summer

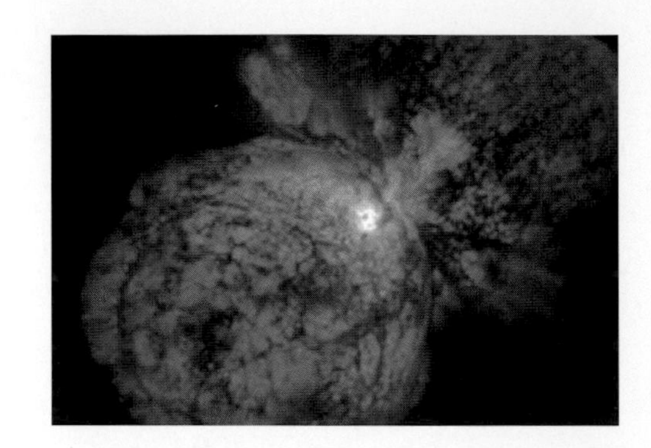

IV. Autumn

Acknowledgments

This book owes its existence first of all to Jacqueline Johnson of Walker & Company who conceived the idea and invited its making. This is the second book I have done with Jackie; every author should have such a gifted editor. Thanks also to publisher George Gibson, design director/managing editor Marlene Tungseth, designer Maura Rosenthal, and the other fine people at Walker & Company, and to my agent, John Williams. Dan Raymo of Platypus Multimedia prepared the star maps. Although I have never met them face-to-face, I owe a debt of gratitude to Guy Ottewell and Fred Schaff for sharing their vast knowledge of the night through their various books and publications. Tom Lorenzin has been a steady friend to both me and the night. Mike Horne shared with me more starry nights than I can count. My wife, Maureen, rolled over many a night to find my side of the bed empty; she knew where I was. My son Tom has fed my enthusiasm for astronomy since he was a toddler. To all of my family, a hearty thanks.

This book is dedicated to the many Stonehill College students over the years with whom I have had the honor of sharing my love of the night, especially John Darak and Chris Asbell who are mentioned herein.

Introduction

WE ARE CHILDREN OF THE NIGHT. Our species evolved under an overarching canopy of stars, undiminished by industrial smog or artificial light. During long hours of darkness there was nothing much to do but sleep and wonder—wonder at the dreamy dots of light that moved overhead in uncountable numbers. Onto this screen of stars our ancestors imposed their hopes and fears. They found suitable candidates for worship among celestial bodies, especially the Sun with its giving and taking of warmth and light. The rains came and went with the stars. The flooding of rivers and fertility of the soil. And there was terror, too. The frighteningly unexpected. Comets. Eclipses. Stars that appeared out of nowhere and then faded away. Within this apparent chaos our ancestors found order. They recorded the movements of Sun, Moon, planets, and stars, and invented geometrical space, linear time and number to account for what they saw. Science and mathematics, as we know them today, got their start in the eastern Mediterranean world of the fifth, fourth, and third centuries B.C.E. at the hands of sky watchers. Our cultural evolution is bound up in so many ways with observations of celestial phenomena that it is hard to imagine what our story might have been had our planet—like Venus—been permanently shrouded in cloud.

Hard to imagine, too, is the *intimacy* your ancestors had with the night. Today, satellite photographs of the Earth from space show the night side of the planet ablaze with artificial light. Those of us who live in or near cities might live and die without ever seeing things that were as familiar to our great-grandparents as the Sun and Moon: the Milky Way, the Pleiades, nebulosities, faint comets, the zodiacal light. We spend our evenings indoors in front of the television or computer monitor, oblivious of the beauty and terror of the celestial abyss.

This book has as its purpose the reestablishment of intimacy with the night. To that end, it offers the *knowledge* of the dome of night that would have been the heritage of every child in an earlier era. It offers, as well, the rich intellectual *tradition* that flows down to us from the earliest sky watchers, through the astrologers and astronomers of ancient civilizations, to Copernicus, Kepler, Galileo, Herschel, Einstein, Hubble, and the many nameless engineers of national space agencies who have sent spacecraft to visit distant worlds. "What makes the desert beautiful is that somewhere it hides a well," wrote the aviator-poet Antoine de Saint-Exupéry. The beauty we see in the sky, too, takes its power from what we don't see—or see only with the mind's eye. Use the star maps of this book to find and learn the planets, stars, and constellations. Then read the essays that connect what you *see* to the extraordinary *unseen* wonders of the universe. Let your imagination break through the apparent dome of night and flow into the realm of supernovas, black holes, galaxies, quasars, and the radiation of the big bang, arriving ultimately at the singular moment of creation, when space, time, matter, and energy flowed from an infinite and mysterious source. We are made of star dust. Our story is bound up with the story of stars that lived and died long before the Sun and Earth were born. Our fate is intimately tied to the fate of the universe.

The star maps are drawn for an observer at a typical midnorthern latitude—40 degrees north, about the latitude of Philadelphia, Chicago, San Francisco, Tokyo, Rome, and Madrid—but they will be suitable for most folks who live in the northern hemisphere. If you live south of latitude 40 degrees, the stars will stand a bit higher on the southern horizon and lower in the north. If you live north of 40 degrees, the opposite will be true. Unless otherwise noted, the seasonal maps are drawn for about 10 o'clock in the evening on January 15, April 15, July 15, and October 15, but you will soon learn to make adjustments east or west for other times or dates. As you stand under the stars, don't neglect the things that are not on the maps. The five naked-eye planets are a "movable feast"; appendix 1 will help you recognize them. You might see "shooting stars" on almost any clear night, but for the best times of the year to view meteors, see appendix 2. And don't fail to watch for satellites moving with a stately elegance against the background of the stars; you will see them during the hours after sunset and the hours before dawn when they catch the rays of the Sun as they move high overhead.

"It is breathtaking simply to be here," wrote the poet Rainer Maria Rilke. Breathtaking to stand under a starry sky and look deeply into the universe of galaxies. Breathtaking to participate in the constantly changing drama of the night sky. Star watching at its best is a total experience, involving all of the senses—the sights, sounds, tastes, smells, and tactile sensations of the night. It is this complete immersion into darkness and light, informed by knowledge, open to mystery, that renews our intimacy with the cosmos.

I. Winter

A crisp, clear winter night. The stars twinkle. It is not an inconstancy of the stars that makes them seem to shimmer so in the winter sky, but instability of the atmosphere. We look up to the stars through a wrap of cold air that shivers and shakes, especially in this turbulent season.

The southern sky is dominated by the powerful figure of **Orion**, the Hunter, one of the most ancient and familiar constellations. Look for the three equally bright stars of the Hunter's belt; there is nothing else like them in the sky. The names of these stars—Alnitak, Alnilam, and Mintaka—are Arabic translations of ancient Greek names and come to us from the time of Europe's "Dark Ages," when astronomy was cultivated by the Arabs in centers of learning such as Baghdad. Many star names start with *Al-*, which is the Arabic prefix for *the.* Alnitak: "the belt."

Orion is a good constellation to practice seeing the colors of stars. **Betelgeuse** *(BET-el-jews),* at the giant's back shoulder, has a reddish hue. **Rigel** *(RYE-jell),* in the forward foot, is bluish. The colors tell us the temperatures of the stars. Red stars are the least hot—as hot as the burner on your electric stove when it glows red. Blue stars are the hottest. Red, orange, yellow, white, blue: least hot to hottest. The Sun is a yellow star, of average temperature.

The Hunter has two canine companions: **Canis Major**, the Big Dog, and **Canis Minor**, the Little Dog. The Big Dog runs at Orion's feet. The more excitable Little Dog jumps at his shoulder. Each constellation is best learned as a pair of stars, one bright, one less bright. **Sirius** *(SEAR-ee-us),* the brighter star in the Big Dog, is the brightest star in Earth's sky. It is blue-white hot, and not so far away as stars go—nine light-years, or 53 trillion miles. The brighter star in the Little Dog is yellow-white **Procyon** *(PRO-see-on).*

If the night is dark, clear, and the bright Moon is not in the sky, you will see the pale river of the **Milky Way** flowing down past Orion's shoulder and between the two Dogs. This is the faint light of a myriad of stars each too faint to see with the unaided eye—a visual hint of the great spiral Galaxy that is our cosmic home.

Canis Minor
Procyon
Orion
Betelgeuse
Rigel
Sirius
Milky Way
Canis Major
Looking South

Our Sun is one of a trillion stars in the **Milky Way Galaxy**, a flat spiraling whirlpool of stars of a variety of temperatures, sizes, and brightnesses, embedded in clouds of interstellar gas and dust. The Sun, with its planets, is about two-thirds of the way out from the center of this vortex, at the inside edge of a spiral arm, and we see the accumulated light of the other stars of the Galaxy as a band of light around our sky, which the ancients called the Milky Way. When we are looking toward Orion—the winter Milky Way—we are looking toward the outside perimeter of the Galaxy. In summer we are looking toward the center of the Galaxy, and the Milky Way is much brighter.

The stars we see with the unaided eye are only the brightest stars in our neighborhood of the Galaxy. Sirius is hotter and brighter than the Sun; Betelgeuse and Rigel are farther from Earth and are among the most luminous stars in the Galaxy. Rigel is fifty times bigger than the Sun and blue hot. Rigel is in its prime of life, but such huge stars do not live long. Their lifetimes are measured in tens of millions of years, whereas midsize stars like the Sun might live 10 billion years. Betelgeuse is a dying star, a **red giant**. Its outer layers have swelled and cooled; if Betelgeuse were located where the Sun is, its surface would lie beyond the orbit of Mars!

The Dog Star, Sirius, has a tiny companion star, invisible to the eye but visible through a telescope. It is sometimes called "the Pup." The Pup is at the last stage of its life. One billion years ago the Pup was a Sun-like star, in the prime of life. As it used up the last of its energy resources, it swelled to become a red giant, then collapsed upon itself to become a superdense, Earth-sized object called a **white dwarf**. It is no longer producing energy and will slowly cool and cease to shine.

Look closely at the faint "star" in the middle of Orion's knee. You may notice it looks fuzzy. A telescope reveals that it is not a single star at all, but many new stars embedded in a glowing cloud of gas and dust out of which they were born. This is the **Great Orion Nebula**, a stellar nursery.

GALACTIC EQUATOR

Betelgeuse

Great
Orion
Nebula

Rigel

Sirius
and "Pup"

LOOKING SOUTH

1. Alpha

In the Beginning

Here is a way to experience the stars as you never have before—and to come as close as you will ever get to the first moment of creation.

Take yourself as far as possible from city lights, to a place where the night is inky black and thick with stars. If you can, turn off all local lights. Make sure the Moon is not in the sky, or at least no more than a slender crescent. A winter or summer night is best, when the Milky Way arches high overhead and the sky is posted with brilliant stars. Two other requirements: solitude and silence. You'll also want an audio CD player and a recording of Joseph Haydn's The Creation oratorio. Lie back comfortably on a deck chair or a blanket, facing up to the stars. Place your finger on the "play" button and close your eyes. Wait a few moments until you are perfectly relaxed, then, *with your eyes still closed*, push "play."

Silence. A C-minor chord, somber, out of nowhere. Followed by fragments of music. Clarinet. Oboe. A trumpet note. A stroke of timpani. A prelude of shadowy notes and thrusting chords, by which Haydn meant to represent the darkness and chaos that preceded the creation of the world. Listen now, eyes closed, as the music descends into hushed silence. Hear the voice of the archangel

A DEEPEST-EVER VIEW OF THE UNIVERSE, SHOWING HUNDREDS OF GALAXIES NEAR THE BEGINNING OF TIME, BILLIONS OF LIGHT-YEARS AWAY.

Raphael: "In the beginning God created the heaven and the earth, and the earth was without form and void, and darkness was upon the face of the deep." The chorus, subdued, barely audible, sings: "And the spirit of God moved upon the face of the waters; and God said: Let there be light!" Then, the voices whispering, once and only once: "And there was light." Open your eyes! A brilliant fortissimo C-major chord! A sunburst of sound. Radiant. Dispelling darkness. A universe blazes into existence, arching from horizon to horizon. Stars. Planets. The luminous river of the Milky Way. As you open your eyes to Haydn's fortissimo chord and to the (almost) forgotten glory of a truly dark starry night, you will feel that you have been a witness to the big bang.

A modern astronomer would recognize in Haydn's music a breathtaking analogue to our contemporary theory of creation. According to our present story, the universe began 15 billion years ago from an infinitely small, infinitely hot seed of energy—what mathematicians call a *singularity*. The singularity was not "somewhere"—the fabric of space-time came into existence *with* the explosion. There was no "before," at least none that we can presently know. Space swelled from the singularity like a balloon inflating from nothing. Time began. During the first trillion-trillion-trillionth of a second, matter and antimatter flickered in and out of existence. The fate of the universe hung precariously in the balance; it might grow, or it might collapse back into nothingness. Suddenly it ballooned to enormous size (after all, we are here), in what cosmologists call the *inflationary epoch*, bringing the first particles of matter—the *quarks*—into existence.

Within one-millionth of a second, the rapid swelling ceased, and the quarks began to combine into protons, neutrons, and electrons.

The universe continued to expand and cool, but now at a more stately pace. Already it was vastly larger than what we are able to observe today. Within a few more minutes, protons and neutrons combined to form the first atomic nuclei — hydrogen and helium—but still the universe was too hot for the nuclei to snag electrons and make atoms. Not until 300,000 years after the beginning did the first atoms appear. Tiny density variations in the gassy universe of hydrogen and helium were accentuated by gravity, which pulled together the first stars, gassy planets, and galaxies. There were not yet any solid-surface Earth-like planets, because there were not yet significant quantities of elements heavier than hydrogen and helium, such as carbon, oxygen, silicon, and iron. (These elements would be cooked up later in the cores of massive stars and distributed to space when stars died explosively at the end of their lives.) Still, within a billion years of the big bang, the universe had begun to look familiar on the largest scale.

Try to imagine all of this as Haydn's music continues. A bright and lilting melody takes up the story. Musical themes coalesce from chaos. Disorder falls away. The mood changes from somber to gay, and the chorus sings a dancelike tune: "A new-created world springs up at God's command." The chorus repeats the phrase again and again, as if it is not a single world that God creates, but a multitude of galaxies and stars. The bass intones, "And God made the firmament." Music leaps and dances into thrilling passages of sound. "By sudden fire the sky is inflam'd," the bass sings. Thunder rolls from the orchestra. Sixteenth notes fill the air as multitudinous as the stars that silver-fleck your dome of night. This stunning consonance between Haydn's music and the universe above your head is no coincidence.

On a visit to England in 1782, Haydn sought out the astronomer William Herschel, world famous as the discoverer of the planet Uranus. Herschel was himself a transplanted German and a musician, an organist and music teacher who became the most prolific astronomer of the eighteenth century. He had moved to England with his sister Caroline (who would also become an accomplished astronomer) in 1772, and four years later he had constructed the six-inch-diameter reflecting telescope with which he found Uranus, the first addition to the five planets known to the ancients. Not only did the discovery of this planet attract the attention of Haydn; it also won Herschel financial support from King George III, which enabled him to become a full-time astronomer and to construct in 1788 a forty-eight-inch-diameter reflector that remained the largest telescope in the world for half a century.

A view of deep space through Herschel's instrument may have inspired Haydn's musical depiction of God's work on the Fourth Day, the creation of the Sun, Moon, and stars. Certainly Haydn admired the astronomer's giant telescope and may have listened to Herschel's ideas about how gravity condensed the cosmos out of chaos. It was Herschel who first supplied evidence (by plotting the locations of thousands of stars in three dimensions) that the Milky Way is a disk-shaped array of millions of stars (today we would say hundreds of billions), and he guessed that many of the nebulous patches of light visible in his telescope were other Milky Ways, other vast swarms of stars—what the philosopher Immanuel Kant had called *island universes*. After a visit to the astronomer in 1786, the novelist Fanny Burney exclaimed that Herschel "has discovered fifteen hundred universes," an extraordinary bit of gush that was not far off the mark. Later, Herschel would back away from the idea of island universes, but not before he had imagined a universe of galaxies and guessed the shape of our own spiral galaxy, which he compared to a "grindstone."

The orchestra ascends now on a crescendo of luminous sound. "In splendor bright is rising now the sun," the tenor sings. "The space immense of th' azure sky, a countless host of radiant orbs adorns." As you lie beneath the ink-dark night with its thousands of visible stars, imagine the countless other stars that Herschel saw through his giant instrument. Also, imagine the host of nebulae that he observed and cataloged, invisible to the unaided eye (although, if it is summer, you might just see a faint glow in the constellation Andromeda that is the central region of the nearest spiral galaxy to our own). Imagine, as you lie beneath the stars listening to the remainder of Haydn's composition, that this multitude of worlds—expanded by modern telescopes to include hundreds of billions of galaxies—was brought into being by that whispered evocation "And there was light," that singular, explosive C-major chord that accompanied the opening of your eyes. Wonder, too, that the big bang universe of galaxies resides within your consciousness. Haydn's Creation oratorio and Herschel's universe of galaxies have become our own, and not even the mystery of the creation itself rivals the greater mystery of how, through the instrumentality of human art and science, the universe has come to know itself.

There is one significant difference between Haydn's libretto, which is based upon Judaic scriptures, and the modern astronomer's story of creation: The explosive chord that follows Haydn's "And there was light"—his musical big bang—has a *prelude*, those minutes of shadowy and slightly discordant sounds that represent the darkness and chaos that preceded "the beginning." The big bang of the astronomers has no prelude, at least none that we can know. The universe begins as a singularity—a mathematical infinity that poses an impenetrable barrier to knowing what came before. There are a few speculative cosmologists who try to tease from their equations some glimpse of what might have preceded the creation of the universe 15 billion years ago—perhaps a frothing foam of universes that bubble in and out of existence like the effervescence of champagne—but all of this must for the time being be considered science fiction. For all practical purposes, the big bang had no "before."

Nevertheless, many striking similarities remain between the creation described musically by Haydn and the astronomer's big bang. These similarities may represent prescient insights on the part of Haydn and the authors of Genesis. More likely they represent conceptual limitations of the human brain.

There are only so many ways a universe can be imagined coming into existence. Myth, art, and science work from the same limited repertoire of metaphors. Haydn's universal architect speaks, and there is light. For the modern astronomer, space and time begin as a fiery explosion from an infinitely small, infinitely dense seed of energy: the big bang. The seed has no human face, but it bears the lineaments of human creativity.

The concept of the big bang had its genesis in a single startling discovery, made in the 1920s, with a big new telescope on Mount Wilson in California. The instrument had a mirror more than twice the diameter of Hershel's largest instrument, and it showed something that Herschel had not been able to see: Many of Herschel's blurry nebulae were composed of vast swarms of individual stars. They were in fact other "island universes," as Herschel had guessed—other Milky Ways of hundreds of billions of Suns, in the shapes of balls, ellipsoids, and, most spectacularly, pinwheel spirals. We now call these star swarms galaxies, from the Greek word *galaktos*, "milk." Because astronomers could now see individual stars in the nebulae, it was possible to estimate their distance by their apparent brightness. The distances turned out to be staggeringly great. The nearest of the spiral nebulae, the one we see with the naked-eye as a blur of light in Andromeda, is 2 million light-years away! None of this was terribly surprising to the astronomers; after all, Herschel had guessed as much more than a century earlier. The big surprise was something different, something discovered by the astronomers Edwin

Hubble and Milton Humason: The galaxies are racing away from us, or, more accurately, racing away from one another. The universe of the galaxies is expanding!

And if the galaxies are moving apart, then they must have been closer together in the past. Theoretically, we can run the movie backward, using the laws of physics to tell us what happens. The galaxies converge. The density of matter increases. The temperature soars. Atoms dissolve into their constituent parts. Mass evaporates into pure energy. Run the movie 15 billion years into the past, and the whole thing—the entire universe of galaxies that exists today—collapses into an infinitely small, infinitely dense, infinitely hot mathematical point. The singularity. The progenitor of the big bang.

How do we know that running the movie backward gives a true picture of the beginning? We do experiments here on Earth with high-energy particle-accelerating machines to figure out how matter behaves at extremely high temperatures. Using this knowledge, we calculate what sort of universe should have emerged from the big bang. Then we compare the calculated universe to the universe we observe with our telescopes. So far the fit is excellent. Two modern observations in particular confirm our confidence in the big bang. Theory predicts that when ordinary matter first condensed from the hot particle soup of the big bang, it should have consisted almost entirely of hydrogen and helium in a ratio of about three to one, and that's exactly the ratio of these elements that we find in the universe today. Also, the theory predicts that if we look far enough out into space, and therefore far enough back into time, we should see the flash of the big bang as a blaze of luminosity that has subsequently cooled into invisible microwave radiation. This prediction is in exact agreement with observations made with a space telescope called Cosmic Background Explorer (COBE).

As I write this, physicists at the Brookhaven National Laboratory on Long Island are cranking up the temperature of their experiments higher than ever before—to one trillion degrees! They will hurl heavy atomic nuclei in opposite directions around a powerful $600-million accelerating machine—the Relativistic Heavy Ion Collider—until the nuclei are moving at nearly the speed of light, and then smash them into one another. Out of these titanic nuclear collisions, which last only the tiniest fraction of a second, they hope to see emerge a new kind of matter—a *quark-gluon plasma*—the presumed superhot primordial broth out of which all ordinary particles were born. No one has seen a naked quark before, or the gluons that supposedly bind the quarks into protons, neutrons, and electrons; they have not existed since the earliest instants of creation. If these new experiments are successful, and the quarks and gluons show themselves, physicists will catch a glimpse of what the universe might have been like in its first millionth of a second, before protons, neutrons, and electrons came into existence. Then they can run the theoretical movie backward even farther into the past and compare the calculated universe and the observed universe with yet more exactitude.

No one could have been more surprised by a big bang beginning than the twentieth-century astronomers themselves. Creation from nothing was not a story they favored. For cultural reasons—perhaps a reaction against the scriptural scenario or the intractability of that mathematical singularity—they preferred the tranquil poise of a universe that had existed forever. They expected no boundaries when they looked out into the universe with their telescopes, no infinities that could not be conceptually climbed, only space and time stretching on and on into the past and future without limit. In 1915, Albert Einstein glimpsed the possibility of a beginning as he played with the equations of his new theory of general relativity; his equations predicted a universe that must expand or contract. But Einstein was so repelled by what he saw—a singular moment of beginning or a catastrophic end—that he added a fudge factor to his theory to make the universe settle down into tractable serenity. He would later say that it was the biggest mistake he ever made, for not long thereafter the astronomers on Mount Wilson showed him his error.

It is a grand adventure, this search for origins. It is what drew Haydn to Herschel's observatory and inspired the composer's magnificent Creation oratorio, whose libretto was drawn from the Judaic scriptures. Scientists of today are doing no more than did the authors of Genesis thousands of years ago: They are inventing stories of the beginning. Unlike the makers of myth, today's storytellers have evolved rigorous experiments and observations to test their inventions in the refining fire of experience.

And there was light! An explosion from an infinitely small, infinitely hot seed of energy that brings into existence 100 billion galaxies. Who can imagine it? *Mathe-*

matical singularities. Quarks and gluons. Cosmic microwave background radiation. Where in all of this is the human face, the consoling visage of a Divine Artificer who speaks in a language we all understand: "Let there be light"? But remember: The story that Haydn adopted for his libretto was invented at a time when people believed the Earth was the center of a compact universe, and that the stars were *just up there* on the dome of night—"night's candles," to use Shakespeare's lovely phrase. Already, in Haydn's time, the evidence of Herschel's telescope had begun to make the old story seem paltry and false. We needed a grander story, a narrative that could encompass island universes of one trillion stars.

Can we ever learn to feel at home in a universe of 100 billion galaxies? Will we ever experience intimacy with the universe of the big bang? Haydn was not afraid of what he saw through Herschel's telescope; what he saw (he surely believed) was the face of God revealed in his creation. He was exhilarated, inspired to soul-stirring art. He did not hesitate to adapt the Genesis story to praise the new universe of stars and nebulae, for he understood that wisdom is where we find it. And now, on this dark and starlit night, as the stars drift toward their westward setting and the music draws toward its conclusion, listen as the voices of Adam and Eve exalt in the world as they find it:

Adam: The morning dew enlivens all!
Eve: The cool of evening all restores!
Adam: How refreshing the sweet sap of ripe fruit!
Eve: How pleasing the fragrance of the flowers!

Now that you know Orion, lift your eyes a bit higher in the sky and see why his club is raised (with his Betelgeuse arm) and why his lion-skin "shield" is held out in front of him. **Taurus** *(TOR-us),* the Bull, is charging from the West, its long horns lowered toward Orion's head. Look especially for a **"vee"** of stars, all of them faint except **Aldebaran** *(al-DEB-a-ran)* in the upper left tip of the vee. I like to think of the vee as the face of the Bull, with Aldebaran its fierce red eye.

Taurus is one of twelve constellations of the **zodiac** that the Sun moves through during its annual journey around the sky. The Sun's apparent path among the stars is called the **ecliptic**; it is along this line that eclipses of the Sun and Moon always occur. Because the solar system is more or less flat, you will always find the Moon and planets somewhere near the ecliptic, within a zodiac constellation. If you see a bright "star" near the ecliptic that is not on the star map, it is almost certainly a planet.

Aldebaran has a reddish hue. Its name means "the Follower." It follows the Pleiades across the sky.

The **Pleiades** *(PLEE-a-deez)* are part of the constellation Taurus. This wonderful little "teacup" of stars is unique in the sky. Some people mistake the Pleiades for the Little Dipper, a less interesting group of stars in the northern sky. How many Pleiades do you see? Most observers see six stars, which raises the question: Why has this little cluster been called the Seven Sisters since antiquity? In Greek myth, the sisters are daughters of the ocean nymph Pleione; hence their name. Later, I will offer one story about what happened to the missing sister.

Raise your eyes to the **zenith**, the point directly overhead. The star that dominates this part of the sky in winter is **Capella,** a yellow star, the fourth-brightest star for northern observers—after Sirius, Arcturus, and Vega. Capella is in the constellation **Auriga** *(or-EYE-ga),* the Charioteer. Look for a pentagon of stars. Capella means "she-goat," and the three faint stars not far away are the **"Kids."** One star of the pentagon is shared with Taurus; it is the tip of one of the Bull's horns.

Capella (near Zenith)
"Kids"
Auriga
Nath
Pleiades
Ecliptic
"Vee"
Aldebaran
Taurus
Orion
Looking South

The stars of the Bull's face, excluding Aldebaran, are part of a true cluster called the **Hyades** *(HI-a-deez)*. Binoculars or a telescope will reveal many more stars than those you see. These stars were born together out of the same nebula of gas and dust, and are now drifting away from the place of their birth. They are the nearest cluster to us, only 130 light-years distant, but their drift is taking them farther away. Millions of years from now they will no longer be visible to the naked eye.

Aldebaran is not part of the cluster but lies closer to us along the same line of sight. It is a red giant star, near the end of its life, although not nearly so large as Betelgeuse.

The Pleiades are a very young cluster of stars, only a few tens of millions of years old—mere infants as stars go. Long-exposure photographs reveal hundreds of stars still swaddled in remnants of the nebula out of which they were born. They are about 400 light-years from Earth. Don't be fooled by the faintness of these stars in our sky; the brightest of the Seven Sisters, **Alcyone** *(al-SIGH-oh-nee)*, is a thousand times more luminous than the Sun.

Not far from the tip of the Bull's lowered horn, and invisible to the naked eye, is the famous **Crab Nebula**, the shattered remains of a massive star that blew up in the year 1054 C.E., Earth time. Since the star was many thousands of light-years from Earth, the explosion actually occurred thousands of years before it was observed by humans. For weeks this **supernova** was the brightest starlike object in the sky, rivaling Venus at its brightest. Sky gazers everywhere must have watched with wonder.

As stars burn, they make heavy elements—such as carbon, nitrogen, oxygen, and iron—out of the lighter elements hydrogen and helium. When stars blow up, they disperse these heavy elements to space, where they might then become part of future stars and planets. When we turn our telescopes toward the Crab Nebula, we see a crab-shaped tangle of gasses racing outward, seeding the galaxy with freshly minted elements of life.

Capella

Crab Nebula

Pleiades

Hyades

Aldebaran

LOOKING SOUTH

2. Dark

Why the Night Is Dark

In a poem titled "He Wishes for the Cloths of Heaven," William Butler Yeats muses:

Had I the heavens' embroidered cloths,
Enwrought with golden and silver light,
The blue and the dim and the dark cloths
Of night and light and the half-light,
I would spread the cloths under your feet.

But he is poor, the poet continues, with only his dreams. So he spreads his dreams beneath his lover's feet, gently urging, "Tread softly because you tread on my dreams." Few more beautiful words have ever been put on paper. Yet the poem never made much sense to me, for surely the one thing that belongs to all of us, rich and poor, are the cloths of heaven. Impoverished poets and billionaires have equal access to the beauty of the starry night. In fact, many of the poorer peoples of the world, in the developing countries, have greater access to the darkness, for they

The Pleiades, a nearby cluster of young stars, shrouded in wisps of the nebula out of which they were born.

have been less able to pollute the night with artificial light. The blue-dim tapestry of night—spread from horizon to horizon, studded with diamond lights, and embroidered with the golden and silver threads of the Milky Way—has inspired religion, myth, mathematics, and science ever since the first sparks of consciousness ignited in human brains. Even today, in our technologically sophisticated times, a view of the night sky from a truly dark place cannot fail to inspire dreams of grandeur and of a meaning greater than ourselves.

We are animals who evolved in a world without artificial light. Our brains were sculpted as much by darkness as by light. The planet turns on its axis, bearing us from the Sun's light, to half-light, to night, to half-light, and back to light—again and again. Nothing, absolutely nothing in our ancestors' lives, not even the changing seasons, was more constant, more certain, than the diurnal immersion into darkness. Out of fear and convenience, they confined their vigorous activities to daylight hours and spent the dark times huddled together in waking and sleeping dreams. Dreams are the fruits of darkness—exhilarating, terrifying glimpses into the labyrinthian chambers of the mind, the source of our most brilliant inspirations and deepest fears.

No wonder poets so love the dark. In the bardic schools of ancient Ireland, young poets in training were assigned a theme by their teacher, then sent to private cells furnished with nothing more than a bed upon which to lie and a peg upon which to hang a cloak, and—most importantly—without windows, not even a crack through which light might enter. There, in total darkness, the young students were expected to compose their rhymes, throughout the night and all of the next day, undistracted by a single ray of the Sun. Then, after a complete cycle of the Earth on its axis, they emerged into the light and wrote down their lines. We remember that only after the poet Dante had made his way through the Dark Wood, with Virgil as his guide, did he see—really see: *E quindi uscimmo a riveder le stelle*. "And so we came forth again and saw the stars."

Our ancestors were star watchers perforce. At night they crept out of their caves or crude shelters, and the glittering dome of night arched from horizon to horizon, filling half of their visual field. What did they make of it, that light-flecked canopy of darkness? Today, we would say that the stars are scattered across the sky essentially at random. But our ancestors were frightened by chaos, and so they imposed familiar images upon the stars: animals and gods, agents of good and evil, tribal totems. The darkness was a screen upon which they projected their hopes and fears and dreams. The dark night was the first book of poetry and the constellations were the poems.

Throughout the night the unchanging patterns of the stars crept unhesitatingly across the sky, a hand span every hour, lifting their lights above the eastern horizon, extinguishing them in the west. In the north or south, the stars went round and round, circling a still point in the sky, the *celestial pole*. As the seasons passed, the patterns of stars that were visible in the evening changed: Orion and his dogs in the winter, the great birds Cygnus and Aquila in the summer. Twice each year the Milky Way swept overhead, a flowing river, a bridge between

Heaven and Earth. Parents took their children beyond the campfire's circle of light and recited the poems of the constellations, and so the poems were passed from generation to generation, and the night became less frightening, more familiar.

"In a dark time, the eye begins to see," wrote the poet Theodore Roethke. It is one of the oldest themes of literature. Isaiah said it too: "He who walks in darkness has seen the light." Every cultural tradition has its nocturnal vigils, its dark nights of the soul. Poetry without darkness is not poetry at all. Yet we seem bent upon surrendering the darkness for a mess of electric light. As I write this, a team of men from the electric company is mounting a streetlight on the utility pole at the end of my driveway, a poorly designed light, certain to send a yellow glow up into the sky where it does no good whatsoever. Progress? Security? I ask the students in my astronomy classes: How many of you have seen the Milky Way? One student in ten raises a hand, those few who have spent some time away from cities and suburbs, in a wilderness camp perhaps, on a sailboat at sea, or in some remote holiday location. It was possibly the Milky Way's faint light, spilling across the darkness, that inspired these lines of the seventeenth-century poet Henry Vaughan:

I saw Eternity the other night,
Like a great Ring of pure and endless light,
All calm, as it was bright;
And round beneath it, Time in hours, days, years,
Driven by the spheres
Like a vast shadow moved; in which the world
And all her rain were hurled.

The dark night is one of nature's most precious gifts, a rare and valuable cloth embroidered with the history of our race, which, to our detriment, we fritter away.

Giordano Bruno was almost a contemporary of Henry Vaughan. Their lives might have overlapped had Bruno's life not been cut short by the Roman Inquisition. He was burned at the stake in 1600 in Rome for a long list of heresies, including the idea that the Earth, orbiting the Sun, is just one of an infinitude of inhabited worlds. The stars are other Suns, he thought, receding into the distance without limit, and inhabitants of a planet of any star might foolishly think they were at the center of the universe.

Bruno was born in the Kingdom of Naples in 1548, only a few years after the death of Copernicus. At the age of twenty-four he was ordained a Dominican priest, although his curious and uninhibited mind had already attracted the disapproval of his teachers. Within a few years of ordination he was accused of heresy. The idea of heresy meant nothing to Bruno; he claimed for himself (and for others) the right the philosophize, to dream, unfettered by authority or tradition. Like the poet Vaughan, he saw in the heavens a ring of pure and endless light, a universe of such infinitude as to terrify his contemporaries. Bruno embraced those infinite worlds with zeal. He was a modern in many qualities of mind: materialist, rationalist, a champion of free and skeptical inquiry. He made no distinction between matter and spirit, body and soul, and yet he was profoundly aware of the "vast shadow" that moves upon the face of the night, in which "the world and all her rain [are] hurled."

Poet, philosopher, loose cannon, Bruno spent most of his life wandering across Europe. Wherever he went—Italy, Geneva, France, England, Germany—he stirred up a fuss, shaking preconceptions, rattling complacencies, asking philosophers and shopkeepers to stop for a moment and entertain a doubt or two. The universe and God might be bigger than we think, he said. In 1591, at the request of a prospective patron, Bruno returned to Italy, to the Republic of Venice, perhaps because he was homesick, or perhaps because he sought the chair of mathematics at the University of Padua, which he knew to be open. It was a big mistake. Soon he was denounced to the Inquisition by his erstwhile patron. He was extradited to Rome, where he languished in a prison of the Holy Office for seven years, struggling to accommodate his tormentors without forsaking his principles. Accommodation proved impossible. In February of 1600 he was taken gagged to the Campo de' Fiori (Field of Flowers) and put to the stake. Some time ago, on a visit to Rome, I made my way to the Campo de' Fiori, now a busy market square in the center of the city, to see the place where Bruno was burned. A melancholy and somewhat sinister statue of the philosopher stands in the square, a dark and brooding presence among the bustle and brilliance of the market, erected by secular humanists in the nineteenth century when the unification of Italy liberated Rome from direct papal rule.

Bruno was a dreamer. His vision of an infinity of worlds was a poet's dream. But within a decade of his execution, the dream came closer to reality. The vacant chair of mathematics at the University of Padua was offered to Galileo Galilei of Florence. In the winter of 1610, Galileo turned the world's first astronomical telescope to the night sky and saw things that would change the world forever: the mountains of the Moon, the moons of Jupiter, the phases of Venus, spots on the Sun, and a myriad of tiny stars that twinkled beyond the limits of human vision. He communicated these extraordinary discoveries to the world in a little book titled *The Starry Messenger*, in which he claimed that the universe might be infinite and contain an infinity of stars (prudently, he did not mention Bruno, although surely he knew of the radical philosopher and his fate). A copy of the book made its way to Prague, where it was read by Johannes Kepler, the most important theoretical astronomer of Galileo's time. Kepler disagreed with his Italian colleague. If the universe were infinite and randomly filled with stars, he said, the entire celestial sphere should blaze with light as brilliant as the face of the Sun. No matter which way we look into space, our line of sight must eventually terminate on a star, just as a person in a wide forest must in any direction eventually see the trunk of a tree. And since the night sky is manifestly *not* as bright as day, the universe cannot be infinite. The universe must be finite and bounded, concluded Kepler, and what we see as black night is the dark wall that encloses the universe. Kepler's argument was compelling, yet the idea of an infinite universe was in the air, and succeeding generations of astronomers struggled to explain the darkness of the night sky. One person who wrestled with the problem was the nineteenth-century German astronomer Heinrich Olbers, and the puzzle has come to be known as Olbers's paradox: *If the universe is infinite, and contains an infinite number of stars, why is the night dark?*

Some scientists suggested that if interstellar space is not empty, gas and dust would absorb the light of distant stars. But if the absorbing matter exists, it would eventually become hot enough to reradiate the energy it absorbs, maintaining the brightness of the sky. The discovery that stars are clumped into galaxies also failed to resolve the paradox; Kepler's argument can be applied to galaxies as well as to stars. In the end, the resolution of the paradox came from a surprising quarter: The universe had a beginning—15 billion years ago, according to the big bang theorists. Because of the finite velocity of light, as we look out into the universe we are looking backward in time. If we see a galaxy 15 billion light-years away, we see that galaxy as it existed 15 billion years ago. We cannot see galaxies more than 15 billion light-years away because there hasn't been enough time for their light to reach us. Even if the universe is infinitely big, the part of it that we can see is finite, and therefore the number of stars and galaxies we can observe is finite. This is the resolution of Olbers's paradox: *The night is dark because the universe is young!*

What would Bruno have made of this news of the universe's beginning—the premier astronomical discovery of the twentieth century? He would certainly have applauded the questing spirit and independence of mind that led astronomers to embrace the big bang, even though such a notion violated their intuition of what the universe should be. To do philosophy, one must first put everything to doubt, said Bruno. Einstein agreed: The most important tool of the scientist is the wastebasket, he said; if the evidence of the telescope is otherwise, then the scientist's own precious theory must be modified.

Bruno might also have enjoyed the fact that the big bang is implicit in an observation that is available to everyone—philosopher and shopkeeper, poet and shepherd. We step away from the campfire, away from the ring of artificial light. Above our heads the stars are scattered like gems on a jeweler's black cloth. Standing in darkness, we are witnesses to the universe's beginning.

Once I had the pleasure of talking about the stars with a group of students under clear dark skies at the Caribbean Marine Research Center in the Bahamas. The center has an island all to itself with just a few generator-driven lights. From where we were standing, far away to the south one could just make out the faint glow of George Town on the island of Exuma, but overhead the night was jet-black and filled with stars, so many stars it was difficult to make out the patterns of the constellations. I pointed out familiar things—Orion, the Pleiades, Polaris, the bowl of the Big Dipper peeking over the horizon. I also showed the students things they had never seen before, things only visible under truly dark skies: the winter Milky Way flowing at Orion's back; the faint Beehive cluster of stars in Cancer; the Double Cluster in Perseus; and the zodiacal light, streaming vertically from the western horizon, sunlight reflected from dust in the inner solar system. Earlier that evening we had watched a thirty-hour-old Moon, startlingly thin, kiss the horizon with its eyelash arc.

These things were commonplace to our ancestors—to Bruno, Galileo, Kepler, even Olbers. Beautiful things. Hints of the majesty and complexity of the universe. As we watched from the dark Bahamian isle, we tried to imagine ourselves whirling on our multiple journeys—on the spinning Earth, orbiting the Sun, in a spiraling galaxy, and racing outward from the big bang. We saw, or imagined that we saw, Vaughan's "Ring of pure and endless light, all calm, as it was bright." We saw, or imagined that we saw, the spheres "in which the world and all her rain were hurled." The ability to "see" such majesty is virtually gone now in many parts of the world. Later in this book I have included the Beehive in Cancer and the Double Cluster in Perseus among "Things to See," but for most of us they should be included under "Things to Imagine." Like many other treasures of the night, they have been washed away in a sea of artificial light.

Looking at the night sky from the environs of a city or suburb is the visual equivalent of listening to a live string quartet outdoors in Times Square at rush hour. Go to the Web site of the International Dark Sky Association (www.darksky.org) and look at satellite photographs of the nighttime Earth from space—all lit up like a Christmas tree. The northeastern United States, in particular, appears in the photographs as a sprawling luminous blotch. Europe and Japan are ablaze. All of this light directed upward has no utility on the ground; it provides no security or convenience for our nighttime activities. It does, however, scatter through the atmosphere and shroud the planet with a baleful glow that obscures the stars. Light pollution is especially troublesome for astronomers. Their ability to peer deeply into the universe is severely compromised by artificial light. The Mount Wilson telescope, which Hubble used to discover the expanding universe, now sits mostly useless on its mountain above Los Angeles, in a haze of artificial light. Increasingly, astronomers flee with their instruments to the last remaining dark sites in the world—the mountains of Chile, the high extinct volcano on the big island of Hawaii. But as the world's population grows, these places too will be impaired. One hundred years from now our major research observatories will be in space or on the Moon.

The International Dark Sky Association tries to educate the public on economically advantageous lighting alternatives that accomplish the required purpose—business, travel, security, aesthetics—while being minimally intrusive where light is not wanted or needed. The association estimates that wasted, upward-directed light costs the United States alone one billion dollars a year. It makes this analogy: If we had a water sprinkler system that wasted much of its water by scattering water everywhere—onto the street, through our neighbor's windows, and upward to encourage evaporation—we'd not tolerate it for long. If we wasted one billion dollars' worth of water this way every year, we'd declare it a national disaster. But more is at stake than money. The wasted light deprives us of something important to the human spirit: Knowledge and experience of the universe that spawned us, the dark night that inspires the mystic and the poet. The students I spoke to at the Caribbean Marine Research Center were aware of the need to keep the sea and atmosphere free of chemical pollutants. However, like most of us, they had not given much thought to one of the most pernicious pollutants of all: wasted light that separates us from the majesty of night.

We live near a brilliant star, the Sun, and we experience darkness only if something blocks the Sun's light. We can of course shut ourselves up in a lightless room like the student bards of Ireland, but what blocks the Sun's light most prominently is the Earth itself. As the Earth turns each twenty-four hours, it carries us again and again into the planet's shadow. We spin into our envelope of darkness—"the blue and the dim and the dark cloths of night"—dreaming in the shadow of the Earth, glad for the light and warmth of the Sun, but blessed too with an unimpeded view of the deep universe. We have squandered night with a surfeit of artificial light. We should protect the darkness that is left. To those fellows from the electric company putting up a new streetlight at the end of my drive, I say, "Tread softly because you tread on my dreams."

What you see in the sky depends on three things: your latitude on Earth; the time of day; and the season of the year. Each hour, as the Earth rotates eastward on its axis, the celestial sphere and all it contains seem to move westward by about a hand span with arm stretched out. Each day, as the Earth circles the Sun, celestial objects drift westward by about a finger's breadth from where we saw them the night before at the same hour.

To the east of Orion and a bit higher in the sky you'll find **Gemini,** the Twins. **Castor** is the more northerly star; **Pollux** is slightly less bright and has a yellow hue. In Greek myth, Castor was a talented horseman and Pollux was a boxer. They were sons of the mortal Leda, seduced on her wedding night by the great god Zeus in the form of a swan. Leda bore two sets of twins, Pollux and Clytemnestra, immortal offspring of Zeus, and Castor and Helen—she whose lovely face launched one thousand ships—mortal offspring of Leda's legitimate husband. Devoted half brothers, Castor and Pollux became revered as protectors of sailors at sea, who swore "by Gemini"—*by jimminy.* With a little imagination, it is not hard to trace out the figures of the brothers, arms around each other's shoulders, their bodies reaching toward Orion.

Gemini is another zodiac constellation; the Sun moves through Gemini on its annual pilgrimage around the sky. Of course, it is actually the Earth that is moving rather than the Sun, but this was unknown to the ancients.

Between Gemini and the prominent constellation Leo is a rather featureless stretch of sky, but since the Sun spends a month traversing this "gap," the constellation that resides here—**Cancer,** the Crab—has assumed an importance out of proportion to the brightness of its stars. You will need a dark night to see any of them! If the night is very dark, look for the **Beehive,** a hazy patch of light just to the west of the two central stars. Other names for the blur: the Little Cloud, and Praesepe *(pree-SEE-pee),* "the manger" from which the two nearby "donkey" stars eat their hay.

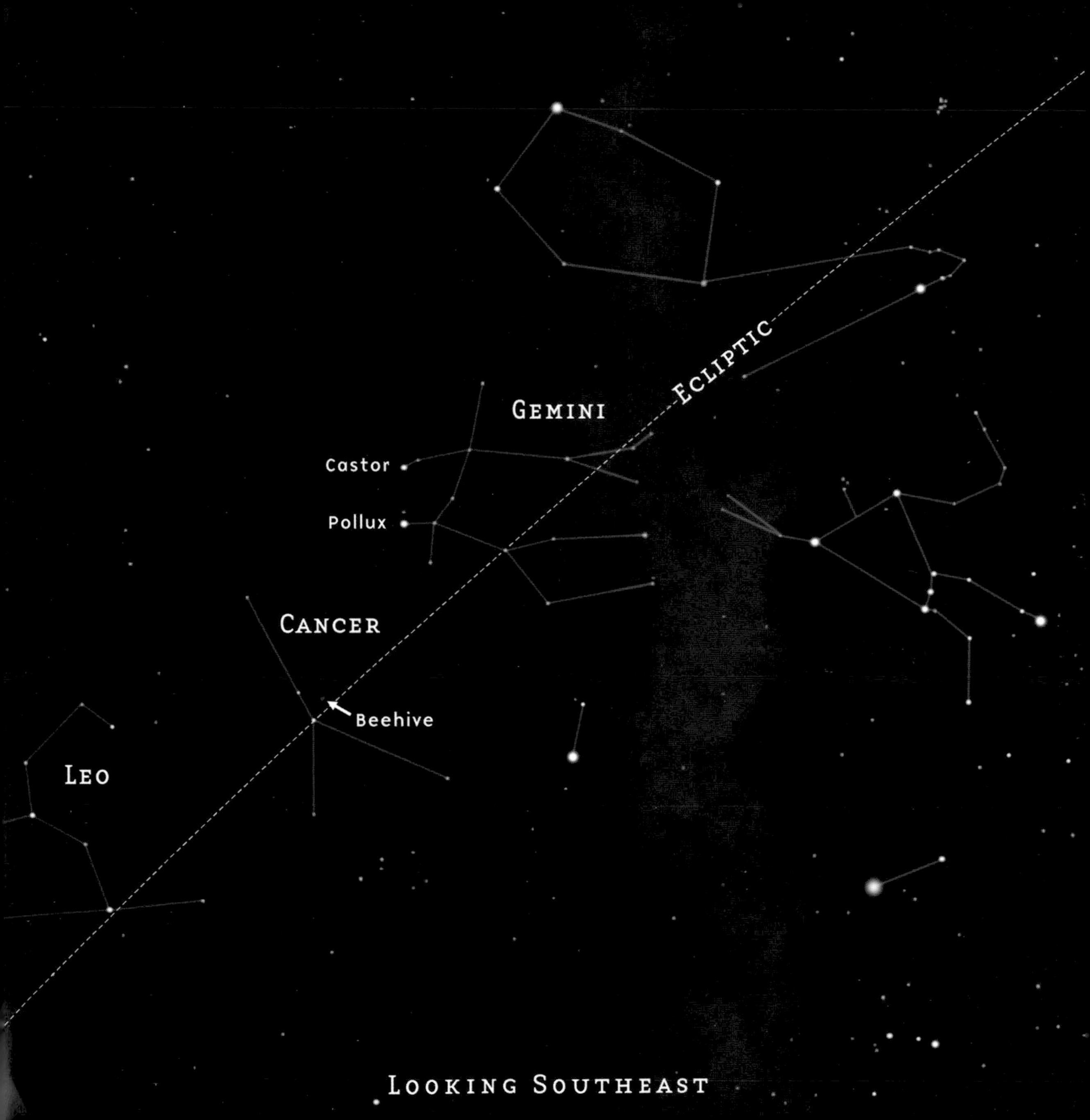

Ecliptic
Gemini
Castor
Pollux
Cancer
Beehive
Leo
Looking Southeast

When Galileo turned his telescope upon the Beehive in 1610, he was astonished to see a cluster of tiny stars, forty of them by his count. With modern instruments we see hundreds of stars. Since the Beehive is so close to the ecliptic, a planet will sometimes visit the "hive," and when this happens the view through binoculars or a small telescope can be quite lovely. Like the Pleiades, the Beehive is a cluster of young stars, although not quite so young as the Pleiades. More than one hundred stars in the Beehive are more intrinsically luminous than the Sun.

The stars in the Beehive were all born at about the same time, pulled together by gravity from the same nebula of gas and dust. It is the *amount of matter* that gets pulled together—the mass—that determines the color and brightness of a star during the prime of its life. The heftiest stars, with masses ten or twenty times the mass of the Sun, burn hot and bright, expending their energy quickly; these blue-white giants live for only tens of millions of years and die violently as supernovas. The least-heavy stars, with masses only one-tenth the mass of the Sun, burn cool and dim; these *red dwarfs* live for tens of billions of years. Since the universe is only about 15 billion years old, every red dwarf that was ever born is still alive.

Among the stars of the Beehive are many similar to the Sun—indeed, virtually identical. Whether they have planets like those in our own solar system we do not know, but we have no reason to doubt it—and every reason to think it might be true. It is likely that our Sun was born as part of a cluster of stars, like the Beehive or the Pleiades, but in the 4.5 billion years since the Sun's birth it has drifted away from its siblings. The stars of the Beehive may themselves disperse as time passes.

Although Castor and Pollux look like twins in the sky, they are actually very different. Pollux, the nearer of the twins, is a swollen star near the end of its life, a yellowish giant. Castor is a complex system of six stars which appear to the eye as one, including three close binaries, all in the prime of life, bound together by gravity, and engaged in an intricate dance about each other.

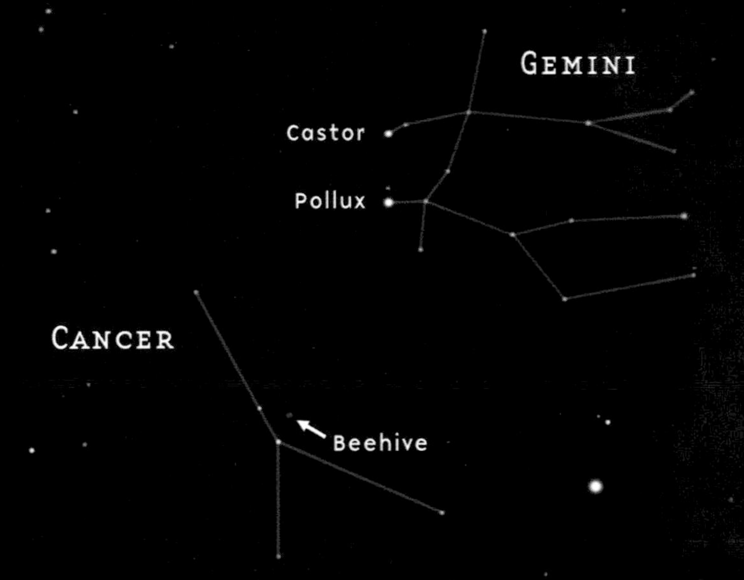

Looking Southeast

3. Far

The Distances to the Stars

I begin my astronomy course at Stonehill College with a sixteen-inch transparent acrylic star globe. The globe shows all the stars that might be seen with the unaided eye from the suburbs of a city on a decent night. Inside the star globe is a smaller terrestrial sphere with a map of the world. Between the two concentric spheres is a "horizon ring" that can be turned and tilted to show the half of the sky that can be seen from any place on the terrestrial globe, and a small yellow ball representing the Sun that can be moved around the sky. I pass the globe around, let my students hold it in their hands, let them feel the tidy *compactness* of it. It is more than just a plaything; for 2,000 years of Western history, this was believed to be an actual image of the world: a central Earth and enclosing sky—a cosmic egg. The idea of concentric spheres was a Greek invention. No gods and goddesses in the celestial realm of the Greek astronomers; just spheres that turn with an elegant mathematical simplicity. And their model of the universe still works amazingly well. We turn the spheres, set the horizon ring, move the Sun

The central and outer region of a globular cluster, a jewel box of brilliant stars, thousands of them, as old as the Galaxy itself.

into its appropriate constellation, and the star globe shows exactly what we will see over our heads at any time of the day or night, at any season of the year, from any place on Earth. A triumph!

I end my astronomy course by projecting onto a big screen in a darkened room one of the Hubble Space Telescope Deep Field Photographs (see photo in chapter 1). To make one of these extraordinary photographs, the Space Telescope focused its camera on a tiny speck of sky—an area that could be covered by the intersection of crossed sewing pins held at arm's length—chosen more or less at random from the apparently dark space between the visible stars. The shutter of the camera was left open for an unprecedented ten days, through hundreds of separate exposures, the longest, deepest looks into space ever contrived. What was revealed in the darkness? A snowstorm of galaxies! Nearly 2,000 galaxies in each tiny speck of sky, including a few relatively nearby spirals and hundreds of faraway galaxies that show up as mere dots of light. The most distant of the galaxies in the photographs are more than 10 billion light-years away. We see them as they were not long after the universe's beginning. It would take 25,000 photographs at this scale to cover just the bowl of the Big Dipper. A similar survey of the entire sky would capture the images of 100 billion galaxies.

The journey we make in my astronomy course from the Earth-centered universe of the star globe to the Hubble Space Telescope's universe of 100 billion galaxies is invariably disconcerting. Psychologically, we still pretty much live in the human-centered, human-scaled universe of the Greek spheres. For most of human history, we imagined ourselves at the center of a universe that was made just for us, illuminated by a sprinkle of stars *just up there* on the dome of night. And now, quite suddenly it would seem, we are set adrift in a blizzard of worlds. My students look at the Hubble Deep Field Photograph. I explain what they are seeing. They nod with understanding. And then one of them will say, "It makes me feel so insignificant." No wonder the Inquisition burned Bruno at the stake when he suggested that the universe might be infinite.

The title of this book is *An Intimate Look at the Night Sky*. But what kind of intimacy can one have with a universe of 100 billion galaxies, each galaxy containing one trillion stars, every star perhaps with planets? There are two ways to answer this question. First, close your eyes for a moment and bring to mind the big bang, the out-rushing snowstorm of galaxies, the seething stars, the whirling planets, everything revealed by the telescopes. It's all there, tucked into the synapses of our brains. *We carry a universe in our heads.* It doesn't get much more intimate than that. Second, the discovery of the universe of the galaxies is a human story, a story of human curiosity, human ingenuity, human courage. There is a little bit of Bruno, Galileo, Herschel, and Hubble in all of us; we are them, they are us. As a blind old man in his seventies, Galileo was forced to kneel on the floor of the Holy Office in Rome and deny that the Earth moves around the Sun rather than residing immobile at the center of the universe. He thereby escaped further harassment, perhaps even Bruno's fate. But his passion for the celestial objects he had contemplated was inextinguishable. According to legend, after his public recantation he whispered under his breath, "And yet it moves." That's intimacy.

Like the proverbial journey of one thousand miles, the journey to the distant galaxies began with a single step. Sometime in the third or fourth centuries B.C.E., a surveyor in the employ of the Egyptian pharaoh, trained to walk with measured pace, stepped out of the central square of Alexandria, at the mouth of the Nile River on the Mediterranean Sea, and walked down along the river to the farthest reach of the kingdom, counting his steps. By the time Eratosthenes of Cyrene turned his attention to figuring out the size of the Earth, in the third century B.C.E., he possessed a map of the Kingdom of Egypt, with distances paced by surveyors.

Eratosthenes was librarian of the great store of scrolls at Alexandria, then the largest library in the world. He made the best maps of the world available at the time and invented a system of longitude and latitude. We know *how* he measured the size of the Earth, but not much else about him. I like to imagine that he was an insatiably curious man, who hung out in the marketplaces and wharves of Alexandria questioning travelers from other parts of the world. "What is the weather like in the place you came from?" "What are the plants and animals?" "What are the languages and customs of the people?" "What do the stars look like?" "How far above the northern horizon stands the polestar?" One day he heard a story about a well in the town of Syene, far down the Nile, in which a person might see the sun reflected on Midsummer Day. Just one story among hundreds. Another story to be tucked away in his growing store of geographic information. Then later, perhaps when he was walking alone or waking suddenly from sleep, it hit him: *That's it! I know how to calculate the size of the Earth!*

How did he do it? If the Sun is reflected in the bottom of the well, then on Midsummer Day the Sun must stand directly above Syene. But Eratosthenes knew from his own observations that the Sun is not directly overhead in Alexandria on Midsummer Day. And he knew why. The Earth is a sphere (he believed), and because of the curvature of its surface the Sun can't be directly overhead two places at once. On Midsummer Day he carefully measured the angle of the Sun's rays at Alexandria. He drew a circle on papyrus to represent the Earth, with radii from the center of the Earth to Alexandria and Syene, and a few more straight lines to represent the Sun's parallel rays at the two cities. He knew the distance from Alexandria to Syene (the surveyor's counted steps). With simple geometry, he calculated the circumference of the Earth—and got it dead-on.

Figuring out the size of the Earth without leaving home was an impressive achievement, but something more important happened when Eratosthenes drew that circle on papyrus. He let a pure mathematical abstraction—a circle—represent our wonderfully complex Earth, with its oceans, clouds, rivers, mountains, vegetation, animals, and human societies. Just a circle scribed with a compass. In that simple turning of the compass, he established the power of mathematical abstraction as an aid to discovery, when linked with exact quantitative observations. Eratosthenes's deduction of the Earth's circumference was the beginning of science as we know it. The key to his success was not trying to understand everything at once. For calculating the size of the Earth, the only thing he needed to think about was the Earth's sphericity. There would be time enough later to tackle the details, if and when he or his successors could figure out how. So this is the lesson we learned from Eratosthenes: *The essence of science is restraint.* Humility in the face of overwhelming complexity. A willingness to say, "I don't yet know." Scientists since Eratosthenes sacrifice omniscience for reliability and leave the world in all of its glorious, ineffable details to the poets and mystics.

Eratosthenes's contemporary, Aristarchus, carried the work a step farther and helped establish the new way of thinking on a firmer footing. Again, we know almost nothing about Aristarchus's life, but we do have his wonderful book *On the Sizes and Distances of the Sun and Moon.* Aristarchus made a few quantitative observations: the angle between the Sun and Moon when the Moon is half full; the duration of a total eclipse of the Moon; the apparent sizes of the Sun and Moon in the sky. Then the papyrus again, the straight edge, and the compass. Circles for the Earth, Sun, and Moon; a long triangle to represent the Earth's shadow; the Moon in the shadow at eclipse. Thanks to Eratosthenes, Aristarchus knew the size of the Earth—one side of one triangle in his diagram. With a bit of geometry—the law of similar triangles—he calculated the sizes and distances from Earth of the Sun and Moon.

What was his result? The Sun, he concluded, was *larger* than the Earth. Much larger! According to Aristarchus's calculations, if the Earth were the size of a grape, the Sun would be a grapefruit twenty feet away! Try to imagine the reaction of his contemporaries. It required weeks of sailing to travel from Alexandria to Greece across the Mediterranean Sea, and the Mediterranean Sea is just a tiny part of the Earth's surface. But you can cover up the Sun with the tip of your little finger held at arm's length. Common sense would suggest that the Earth is the larger object. An illusion, said Aristarchus. The Sun appears small because of its enormous distance. A few of Aristarchus's mathematically adept colleagues might have appreciated his calculations, and perhaps these few acknowledged the possibility that he might be right. But most folks, surely, if they knew about Aristarchus's work at all, must have thought him crazy. There was a terrifying spaciousness to his universe. ("It makes me feel so insignificant," they might have said.) Aristarchus even went so far as to imagine that the smaller object, the Earth, went around the larger object, the Sun, rather than the other way around. The world was not yet ready for such radical ideas.

But Aristarchus's ideas were not lost. A few brave souls improved upon his observations, and every time they did so the cosmos got bigger. By the time of Copernicus, 1,800 years later, the Sun was not a grapefruit compared to the Earth's grape; rather, it was a ten-foot sphere compared to the grape, and one thousand feet away! This very uncommonsensical fact is no longer in doubt. Today we bounce radar beams off the Moon and planets. We have sent spacecraft on journeys around the Sun. We know the dimensions of the solar system to a high degree of precision, and we know that the Alexandrian astronomers got it mostly right. With no tools but a straight edge, compass, and a few sticks for measuring angles, they shattered the dome of night and sent the Earth flying through a deep and bottomless abyss of space. But wait! I neglected to mention their most important tool. Our most intimate possession. The human brain.

A surveyor walked off the distance from Alexandria to Syene, and Eratosthenes used this as a baseline for measuring the size of the Earth. Aristarchus used the size of the Earth as a baseline for measuring the distance to the Sun. And once Copernicus, Kepler, and Galileo had convinced us that the Earth goes around the Sun (as Aristarchus guessed), finding the distance to the nearest stars was only a step away. Here's how it's done. Hold your finger out in front of your nose. Look at it first with one eye, then the other. Notice how your finger seems to move against the distant background. This apparent motion, called *parallax*, depends on the distance of your finger from your face: The farther away you hold your finger, the smaller the apparent shift. We should see a similar shift in the apparent position of nearby stars against the background of more distant stars when we observe them from different positions in the Earth's annual journey around the Sun. And the size of the shift tells us the distance to the stars. The Alexandrian astronomers certainly looked for such a shift, but didn't find one.

The successors of Copernicus also looked for a shift, unsuccessfully. There were only two possible explanations for the absence of stellar parallax: (1) the Earth doesn't move; or (2) the stars are so far away that the shift is immeasurably small. After Galileo, only the second explanation was an acceptable option. So astronomers kept looking. Herschel was looking for stellar parallax when he discovered Uranus. As telescopes improved, it became possible to measure the positions of stars with increasing precision, but still they refused to show the telltale shift. Finally, at the hands of the German astronomer Friedrich Wilhelm Bessel in the 1830s, it became possible to measure star positions to an accuracy of a few 100,000ths of a degree (the angular size of a dime at a distance of sixty miles), and this was sufficient to detect the parallax of the nearest stars. Alpha Centauri, the nearest star, turned out to be 26 trillion miles away! If the Earth were the size of a grape, Alpha Centauri would be a ten-foot sphere (or two spheres, actually, since it is a binary star) 50,000 miles away! Bruno was right: The stars are other Suns blazing with solar glory, but quenched to starry dimness by vast distances.

Step by step. The distance from the Earth to the Sun became for Bessel the baseline for measuring the distance to the stars. One layer of abstraction added on top of another, slowly building up a reliable picture of the world. During the years 1989–93 the extraordinary Earth-orbiting telescope Hipparcos (named for another of the Alexandrian astronomers) measured the positions and parallaxes of more than one million stars with unprecedented precision—ten times more accurately than can be achieved with ground-based instruments. This vast wealth of data is available to everyone on the Internet—the positions of one million stars scattered through space. And these are only the stars within our own little neighborhood of the Milky Way Galaxy, which contains one trillion stars. The journey to the big bang continues. Knowing the distances to nearby stars, we can determine their true brightnesses, and these then become the basis for guessing the distance to stars beyond our galactic neighborhood, even to stars in other galaxies. And so it goes. Step by step. The long walk that began with a single pace from the market square at Alexandria takes us ever deeper into space and time.

Yes, it is a little frightening, that yawning celestial abyss. The Earth—our abode in the universe—becomes like a dust mote in a vast cathedral. One can understand the reaction of Aristarchus's contemporaries or Bruno's contemporaries. It has all happened so fast that our brains haven't had a chance to adapt to this new information. We are whirled with vertigo, skeptical, disenchanted. We feel more at home with the stars described by the poet Gerard Manley Hopkins:

Look at the stars! Look, look up at the skies!
O look at all the fire-folk sitting in the air!
The bright boroughs, the quivering citadels there!
The dim woods quick with diamond wells; the elf-eyes!

Two stumbling blocks prevent us from psychologically embracing the universe of the galaxies. The first can be called the Argument from Human Design: the idea that the universe was created by a Divine Being who thinks more or less the way we think, who shares our human sense of scale, our human sense of what is *appropriate*. Some version of this stumbling block has bedeviled scientific thinking from Aristarchus, to Herschel, to Einstein. Again and again, however, our unwarranted intuition of the Creator's thoughts has proved an unreliable guide in our search for reality. A second stumbling block is what the biologist Richard Dawkins calls the Argument from Personal Incredulity: If it seems impossible *to me*, it must be wrong. The search for the scale of the universe has been a long battle against personal incredulity; at every step along the way the universe turned out to be bigger than anyone thought possible. The modern astronomical distance scale is at once a rebuke to the limitations of our imaginations and a tribute to the power of the boldest, most daring thinkers to transcend those limitations. "And yet it moves," whispered Galileo, and with that (perhaps apocryphal) remark he brought the dreamy universe down to Earth.

II. Spring

The season turns. New stars rise in the east. Old friends slip into the sunset. The sky to the south these spring evenings is unspectacular. The Milky Way, with its generous sprinkling of brilliant stars, has dropped to the southwestern horizon. Now, we are looking straight up out of the disk of the Galaxy; only one thousand light-years in this direction and we are out of the spiral, into intergalactic space. All the stars you see in this direction—Regulus, Spica, and the rest—are closer than that.

The most prominent constellation in the southern sky is **Leo,** the Lion, reclining along the ecliptic, another zodiac "sign." The Sun comes this way in the fall; but now, in spring, the Sun is in the opposite part of the sky and we have Leo in the evening. Look for a backward question mark with the bright star Regulus as the dot. It is easy to make out the figure of a reclining lion, with its front legs stretched out before it. The star at the tail-end of the Lion is **Denebola** *(den-EB-o-la),* "the Lion's tail." We'll see another *deneb* (tail) in the summer sky.

Regulus means "little king"; the ancients considered it one of the royal stars that rule the heavens. Certainly it rules this part of the sky. Some authorities believe that the Egyptian Sphinx, the lion-shaped monument by the Nile with a human face, was meant to associate Egyptian kings with the power of the celestial Lion.

Lonely **Spica** *(SPY-ka),* in Virgo, resides in an even emptier part of the sky. On the next pair of star maps, I will show you an easy way to find it, using the stars of the Big Dipper. The Sun passes this way in autumn, which perhaps accounts for the long association of the Virgin with the goddess of the harvest. On ancient star maps she is often shown with a sheath of wheat in her hand. The stars of the constellation—except for bluish Spica—are so inconspicuous that it is hardly worth trying to make out the figure of a maiden.

Regulus and Spica lie close to the ecliptic, so planets and the Moon sometimes come their way. Occasionally, the Moon will pass directly over one of these stars, an exciting event called an **occultation.**

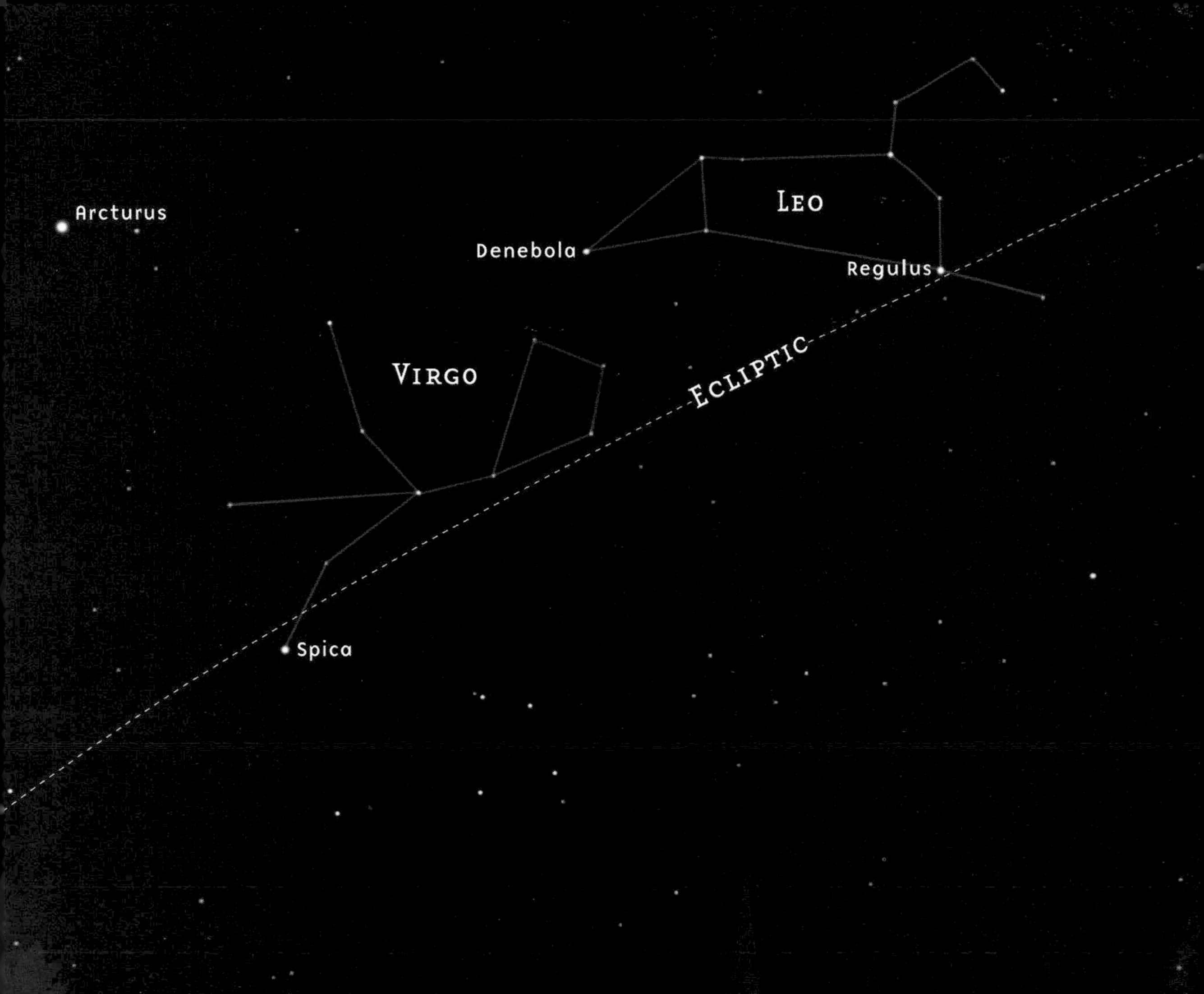

Looking South

When we look in the directions of Leo and Virgo, we are looking up out of the Milky Way Galaxy, with a minimum of galactic dust and gas to block our view of the deeper universe. What do we see out there beyond the stars of the Milky Way? Other galaxies. Other spiraling pinwheels of stars. Other island universes.

Two impressive clusters of galaxies lie close to the hindquarters of the Lion. Several of these galaxies can be seen as hazy blurs with a small telescope; on photographic plates from major observatories the blurs are resolved into gorgeous spirals. Each spiral contains hundreds of billions of stars, most of them, perhaps, with families of planets.

Another noteworthy cluster of galaxies resides in Virgo, on the border of the faint constellation **Coma Berenices** (Berenice's Hair). The **Virgo Galaxy Cluster** contains at least 2,500 galaxies, including many lovely spirals. They lie about 80 million light-years from the Milky Way Galaxy, and they are racing away from us as part of the expansion of the universe. What holds the group together? The amount of mass we see as luminous stars and glowing gas is not enough to gravitationally bind the galaxies together; they should have long since drifted apart. Astronomers speculate that these galaxies—and possibly all galaxies, including our own—are surrounded by halos of unseen mass—known as **dark matter**—that makes up as much as 95 percent of the *stuff* of the universe. What is this nonluminous matter that holds galactic clusters together? No one knows.

Eighty million light-years may seem an unimaginable distance, but the Virgo Galaxy Cluster is relatively nearby as galaxies go. As we look deeper into space, we are also looking back in time. Because of the finite velocity of light, when we photograph a galaxy 10 billion light-years away, we are seeing it as it was 10 billion years ago. Those faraway regions of the early universe are the realm of the **quasars**, mysterious superbright objects that many astronomers believe are associated with the birth of galaxies. The nearest and visually brightest of the quasars, 3C 273 (3C for Third Cambridge Catalogue of Radio Sources) in Virgo, is 3 billion light-years away! Quasars radiate more light than an entire galaxy.

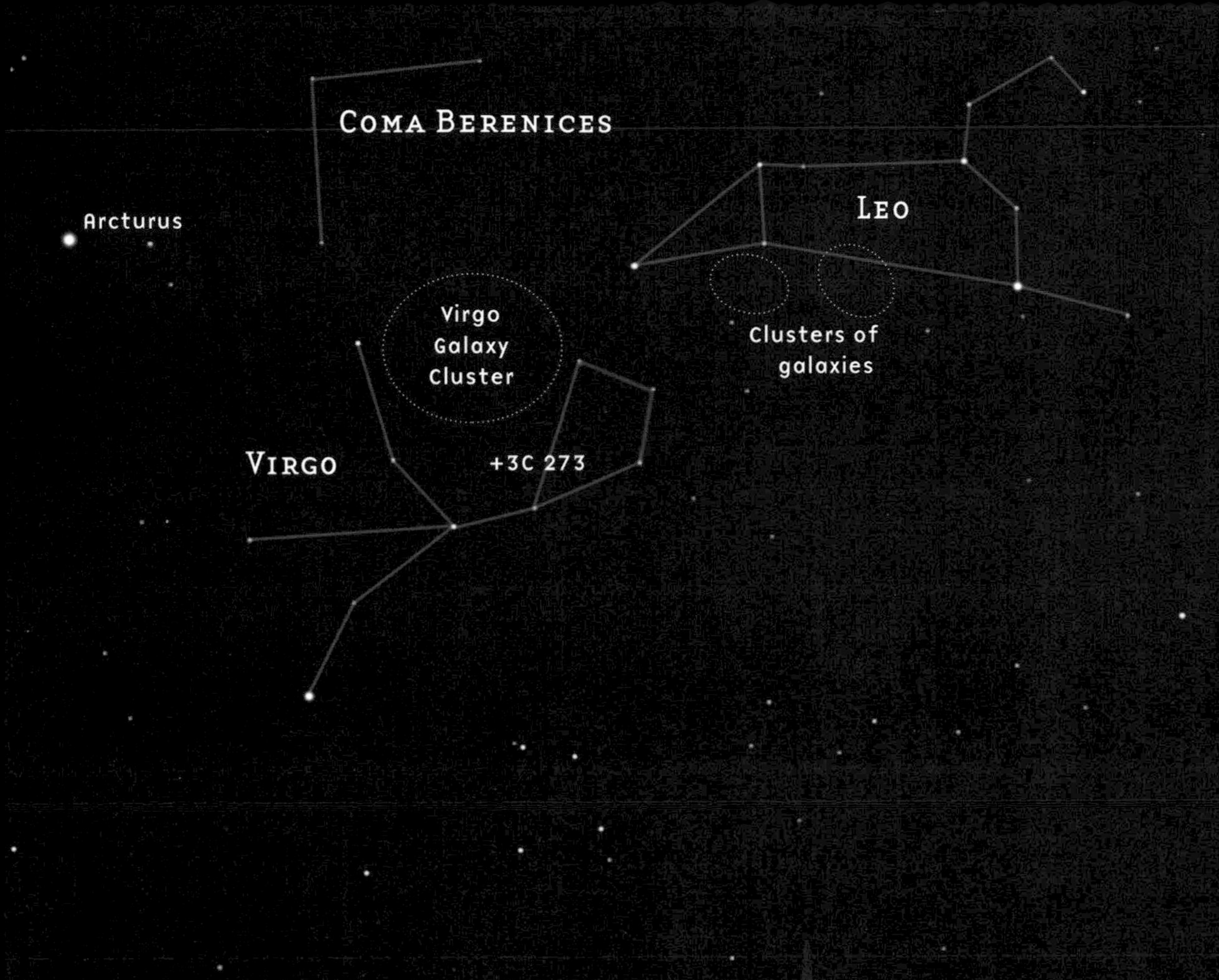

Looking South

4. Fusion

How Stars Burn

Fire-folk sitting in the air. Diamond wells. Elf eyes. The poet's images of the stars beckon us into the forest of the night, away from the security of the campfire or the porch light, into the darkness where 10,000 silver flecks are scattered across the black celestial cloths. We look up into the darkness as if into a mirror and see, as often as not, a human face: Orion, Cassiopeia, Perseus, Andromeda. The human mind has a powerful inclination to see the familiar in the unfamiliar, perhaps related to the tendency of the newborn child to instinctively fix on the mother's face. What we *do not* want to see is the yawning inhuman chasm, the fearful solitude, the infinite multiplicity of worlds. However, what we *wish for* and what we *know to be true* are two different things, and Bruno and Galileo paid the price when their contemporaries failed to make the distinction. Galileo turned his telescope to the flowing river of the Milky Way and saw a myriad of faint lights glittering beyond the limits of human vision. Those uncountable suns were the first hint that the universe was not designed for human eyes.

THE AFTERMATH OF A FLARE-UP ON THE SUN, AS EJECTED MATTER RAINS BACK ONTO THE SURFACE ALONG LINES OF MAGNETIC FORCE.

"The power of the visible is the invisible," said the poet Marianne Moore. The stars are other Suns, other fiery globes vastly larger than the Earth. All of this was known with confidence by the end of the nineteenth century. The distances to the stars had been measured, their true brightnesses determined. The Sun, it turned out, is quite an ordinary star—middling sized, middling temperature, a one-million-mile-wide sphere of gas, yellow hot, seething and churning with energy. Some stars are bigger and hotter than the Sun; some are smaller, cooler, less bright. Furthermore, by century's end, geologists and biologists had determined that the Earth was tens of millions, perhaps hundreds of millions, of years old, and that its surface environment had not changed dramatically during all that time. Among the oldest geologic formations on Earth are rocks composed of marine sediments laid down on the floors of ancient oceans. If the Sun had been hotter during the planet's youth, the oceans would have boiled away; if cooler, surface water would have frozen solid. The constancy of the fossil record also suggests a stable habitat for life. No matter how you thought about it, the evidence was unmistakable: The Sun has been burning rather steadily ever since the planet's birth, pouring into space colossal quantities of heat and light.

And the problem—the biggest problem in science at the end of the nineteenth century—was this: No one had a clue about the source of the Sun's energy. If the Sun was nothing but coal and oxygen, it would have long since burned to embers. Moreover, the Sun was manifestly *not* made of coal and oxygen; the light from the Sun has spectral characteristics of hydrogen and helium. Helium is an inert gas; no energy available there. Hydrogen? Chemists could think of no way to get energy from hydrogen in the absence of appreciable quantities of other elements. What about gravity? Could the squeeze of gravity keep the Sun hot? Physicists did the calculations; if gravity were squeezing out energy at the Sun's present rate of energy production, the Sun would have shrunk noticeably even in historic times. The bottom line: No source of energy known to late-nineteenth-century scientists was adequate to explain starlight.

One evening in 1902, Marie and Pierre Curie put their children to bed and slipped away from their house in Paris. They walked through the still-busy streets to their laboratory, where for four years Marie had labored mightily to extract a fraction of a gram of a new element from tons of pitchblende ore from Czechoslovakia purchased with their meager savings. No government grants were available in those days for Curie's research; no graduate students were there to help her with the backbreaking work. She rolled up her sleeves and spent her precious savings with the purest of motives: curiosity. She was convinced that an undiscovered element was hidden in those heaps of ore, like a needle in a haystack, and she was determined to find it. As she labored, she had wondered what the element she sought would look like. Would it be gray? Black? Shiny? Dull? The answer was unlike anything she might have guessed. As husband and wife entered the laboratory, Marie whispered, "Don't light the lamps." She sat in a chair before the table that bore a sample of the new element in its glass container. The precious element glowed with a spontaneous light! "Look, look," Marie murmured ecstatically, as she gazed at the pale-blue glimmering. Her daughter Eve later called it "the evening of the glowworms."

Marie Curie did not know the source of the mysterious light that came from the stuff she named radium (from its radiance). Here was apparently inert matter that shed an apparently inexhaustible light. Soon her new element was all the rage, and Marie Curie was famous. People daubed their bodies with radium salts and danced like fireflies in darkened nightclubs. They sipped radium-laced cocktails that glowed in the dark. In clock factories, hundreds of young women painted radium salts onto the numerals of dials so that time could be read in the dark, moistening the tips of their brushes with their tongues. Such beauty! Nature's fuelless fire, undying light. As Marie Curie sat with her husband in the darkened laboratory on that evening in 1902, she could not have dreamed that the light from radium had the same source as the light of the stars. Or that the beautiful glow that emerged from the glass container was accompanied by deadly invisible radiations.

The next piece of the puzzle fell into place in 1905. Albert Einstein, twenty-six years old, in Zurich, published three papers that shook physics to its foundations and won him a reputation as a whiz kid of European physics. One of those papers, titled "On the Electrodynamics of Moving Bodies," has been called the most famous scientific paper in history. It was the beginning of the theories of special and general relativity that Einstein elucidated over the following decade.

A supplement to this paper stated for the first time the equivalence of matter and energy, expressed in the famous equation $E=mc^2$ (energy equals mass times the speed of light squared). Einstein wondered if this might be the source of radium's pale light, and his guess turned out to be accurate: In the spontaneous fissioning (breaking apart) of radium nuclei, mass is converted into pure luminous energy. The key to the apparently inexhaustible nature of nuclear energy lies in the factor c^2. The velocity of light is a very large number—186,000 miles per second; squared, an even larger number. A tiny bit of vanished mass yields a lot of energy. The radium in Marie Curie's glass vials glows, and glows, and glows, and the diminishment of its mass is undetectable.

No sooner had Einstein published his equation than astronomers began to wonder if the Sun's prodigious output of heat and light might be due to a conversion of matter into energy. For light elements, such as hydrogen and helium, energy is released when nuclei fuse. Perhaps in the fiery furnace of a star's interior, hydrogen nuclei are fused into helium nuclei in a reaction that converts matter to energy. Perhaps that factor c^2 keeps the Sun shining steadily for hundreds of millions of years. But first, a lot of work needed to be done to confirm this intuition. Experiments had to be performed to demonstrate the transmutation of elements and the consequent diminishment of mass. Marie Curie's daughter and son-in-law, Irene and Frédéric Joliet-Curie, played an important part in these investigations. In January 1934, as Marie lay dying of radiation-induced leukemia, her fingertips crumbling away from radiation poisoning, Irene and Frédéric showed her the first experimental evidence for the artificial production of radioactive energy. Before the end of the decade, the riddle of the stars had been solved.

No stars visible to the naked eye tease us more with the mystery of stellar life than the Pleiades, the unique cluster of faint stars we watched in the winter sky. If these stars were scattered at random across the sky, no one would take notice. They are together because they are young—only tens of millions of years old—and together they attract universal attention. The six stars you see with your unaided eye are only the brightest stars in the cluster. Binoculars reveal dozens of stars; telescopes show hundreds. Long-exposure photographs reveal an even more dazzling view: The brightest of the Pleiades are tangled in wisps of glowing gas. The entire gathering of stars is nested in a faint nebula, made incandescent by the light of the hottest blue stars. We see them still entangled in the swaddling clothes of their origin, a gassy nebula that gravity has gathered into a sisterhood of compact spheres.

The gassy environs of the Pleiades is the nearest of thousands of star-birthing regions in the arms of the Milky Way Galaxy. The cluster is 400 light-years distant, which places it in our own neighborhood of the Galaxy, but still far enough away to reduce the brightest Pleiades to near naked-eye invisibility. (If the cluster were as close as Sirius, we could read at night by the light of its hot blue stars.) If one had looked into this part of the sky 50 million years ago, one would have seen not stars but only a cloud of gas. The gas was mostly molecular hydrogen, although helium was prominent, and there were trace amounts of many other molecules—altogether thousands of molecules per cubic centimeter. This gas was a million billion times less dense than the air we breathe, but within the vast expanse of the cloud there was sufficient matter to make one thousand suns. And always, as everywhere in the universe, gravity was struggling to pull the stuff of the cloud together.

Let's watch as a star is born. A dense knot of gas within the cloud—with a mass of one billion billion billion tons—begins to collapse, squeezed by gravity into an ever smaller sphere. The temperature of the sphere soars, especially at its center. Molecules of paired hydrogen atoms disengage, and then, as the temperature rises, the components of the hydrogen atoms—the proton and electron—are ripped apart. The core of the newly forming star is a blazing sea of protons and electrons. Protons have positive electric charge, and like charges repel. Within the sea of darting particles, the protons keep their distance from one another, approaching and dashing away like bees in a swarm. (The electrons too are dashing about on their own.) But as gravity squeezes further and the temperature continues to climb, the darting protons, moving ever more speedily, get closer and closer before they are mutually repelled. As the temperature at the core reaches 10 million degrees, protons approach near enough for a powerful but short-range nuclear force to come into play. Instead of being repelled, protons are pulled together in a powerful nuclear grip. One proton flings off its positive charge and turns into a neutron (and also sheds a ghostlike particle called a neutrino; more about this later), so that soon the core of the aborning star is abuzz with proton-neutron pairs. Two proton-neutron pairs are welded together by the nuclear force into a helium nucleus, with four nuclear particles. The bottom line: Four hydrogen nuclei are fused into a single nucleus of helium. And here's the wonderful thing. The helium nucleus weighs about 1 percent less than the total weight of the four particles out of which it was made. Mass has disappeared from the universe! *Stuff* has vanished. In its place: pure energy.

A star is born. The pressure of newly released energy pushes out from the core, resisting the crush of gravity. The contraction stops with gravity and fusion in exquisite balance. The star regulates its own burning like the thermostat on your furnace. If gravity gets the upper hand, the temperature at the core rises, releasing more energy to push outward. If fusion gets the upper hand, the star expands, diminishing the squeeze of gravity and lowering the temperature at the core. Every second at the center of a Sun-like star, 660 million tons of hydrogen are converted into 655 million tons of helium by the process known as *thermonuclear fusion*. The missing 5 million tons of mass is turned into an amount of energy equal to the missing mass times the speed of light squared. The rate of conversion is prodigious, but the amount of hydrogen in a star is virtually inexhaustible. A star like the Sun can fuse hydrogen for 10 billion years—until it has a core of helium and death begins. The riddle of the stars is answered. The light of the stars is nuclear energy, the same light that Marie Curie observed in her darkened laboratory on "the evening of the glowworms."

The gassy nebula that gave birth to the Pleiades was gathered by gravity into hundreds of stars. Some of those stars, the hot blue giants, are many times more massive than the Sun; to resist the greater crush of gravity, these stars must fuse hydrogen at a rate that will exhaust their core fuel in only tens of millions of years. The smallest stars in the cluster, the red dwarfs, are less massive than the Sun and need barely fuse hydrogen at all to hold up their own weight; these stars will live for 100 billion years. The hottest, brightest stars in the Pleiades will die explosively sometime within the next millions of years. (None have died yet, or we would surely have observed the shatters.) The coolest stars will live virtually forever. Meanwhile, the Pleiades will drift apart. The cluster will disperse. The longer-lived stars of the cluster will take their inconspicuous places among the trillion denizens of our galaxy. Our own Sun may have once been part of a newborn cluster like the Pleiades. If so, it has long since left the nest and abandoned its sibling stars.

The energy produced by fusion at the center of the Sun requires several million years to make its way up through the half-million miles of overlying gas, absorbed and reemitted many times along the way. At the surface it is released as heat and light, which zips across the 93 million miles of space between Sun and Earth in eight minutes. By the time the energy reaches Earth, all evidence of its source has been erased. But there is a way we can look directly into the center of the Sun, and see what is happening there almost as it happens. The key lies in the insubstantial neutrino (which, as you'll recall, is released as two protons fuse and one turns into a neutron.) Neutrinos are strange little particles. They are about as close to being nothing as something can be and still be something. They have no charge and possibly no mass. However, they carry energy, and they zip along at the speed of light. Because they are wisps of almost nothingness, they interact with ordinary matter hardly at all. Most of the neutrinos created at the center of the Sun fly straight up through the Sun's overlying layers. Within seconds of their birth they are launched into space, and eight minutes later some of them collide with the Earth. Go outside and face the Sun (don't look at it directly!). Feel the heat on your skin. Sense the light on your closed eyelids. The heat and light were created at the Sun's core millions of years ago. But as you stand there in the sunlight, trillions of neutrinos fly through your body every second, neutrinos created at the center of the Sun only eight minutes earlier. You are utterly transparent to these tenuous particles; they stream through you unimpeded. For that matter, the planet Earth is transparent; at night the solar neutrinos fall upon the other side of the planet, whisk right through the body of the Earth, and penetrate your body from underneath the bed! Can you believe this? Can you comprehend that every second of our lives we are blown through with a wind of neutrinos from the Sun? What was that line of Marianne Moore's? "The power of the visible is the invisible."

Neutrinos are hard to catch, but not impossible. At several places on Earth—

in the United States, Canada, Japan, and Italy—vast underground chambers of pure water have been set up to catch neutrinos. Every now and then a neutrino hits a proton or neutron in a water molecule and is absorbed, releasing a tiny flash of light. The pitch-dark chambers are watched with thousands of sensitive electronic eyes, and the flashes of faint light are recorded by computers. As strange as it seems, these underground chambers are telescopes that look into the heart of the Sun, seeing within minutes what is happening there. So far, the number of neutrinos snagged by the detectors is less than what theoretical astronomers have predicted. No one knows why. Perhaps some part of our theory of what is happening at the center of the Sun is wrong, or perhaps we don't yet fully understand the nature of neutrinos and how they interact with matter. The scientists are busy; the answer will come. Meanwhile, they know that their neutrino telescopes are working. In 1987, a star blew up in the southern sky, a supernova, the first relatively nearby supernova in centuries. Briefly it blazed with a glorious light. At the same time that the supernova's light reached Earth, the first large neutrino detectors (in a cavern under Lake Erie and in Japan) sparked with flashes of light. Across tens of thousands of light-years, these elusive particles carried the signature of a dying star.

The atomic nucleus was intensely studied by physicists during the early part of the twentieth century. They knew they were chasing the answer to the biggest question in science: Why is the universe filled with light? By the spring of 1939, the nuclear fusion reaction that makes stars shine had been identified by Hans Bethe, Carl von Weizsächer, Charles Critchfield and others. It was also recognized that when the nuclei of certain atoms of uranium and plutonium fission, neutrons are emitted that can collide with other uranium and plutonium nuclei and cause them to split. Under the proper circumstances, a chain reaction appeared to be possible—a vast amount of energy suddenly released from a small amount of matter.

American physicists asked Albert Einstein to write a letter to President Franklin Delano Roosevelt describing recent fission studies of Enrico Fermi and Leo Szilard. "It is conceivable," Einstein wrote, "that extremely powerful bombs of this type may be constructed." Roosevelt reacted decisively. The Manhattan Project was initiated to produce sufficient quantities of uranium and plutonium to make a nuclear weapon. At dawn on the morning of July 16, 1945, in the New Mexico desert, the terrestrial nuclear age was born. Marie Curie's pale glow was turned into a blaze of light like one thousand Suns. A seething mushroom cloud rose heavenward from the desert floor. Within months, the Japanese cities of Hiroshima and Nagasaki had been obliterated and a global war ended.

A chain-reaction fission bomb—the type of bomb that destroyed the two Japanese cities, will explode spontaneously if a so-called critical mass is assembled. You will recall that it takes extremely high temperatures (such as those at the centers of stars) to fuse hydrogen nuclei into helium, because of the mutual repulsion of protons. Fission bombs can provide Sun-like temperatures here on Earth. It wasn't long before physicists used a fission bomb to fuse hydrogen. On November 1, 1952, a hydrogen fusion bomb was exploded at Eniwetok atoll in the Pacific Ocean; this bomb was many times more powerful than the weapons that obliterated Hiroshima and Nagasaki. The power of the stars was brought down to Earth.

The mushroom clouds of those terrible explosions overarched the twentieth century, casting a shadow over exciting intellectual developments of the first half of the century and nearly scaring us to death during the second half. The latter gloom was somewhat ameliorated by the promise of peaceable uses of nuclear energy. ("The Atom: Our Obedient Servant" was the title of one magazine article in the 1950s.) But the bright promise of harnessing the atomic nucleus to provide cheap, inexhaustible energy mostly evaporated with the accidents at Three Mile Island and Chernobyl. And the collapse of the Soviet Union diminished the likelihood of global thermonuclear war. Many nations still have the capacity to cause nuclear devastation, but all in all nuclear physics is not so much on our minds as we begin a new century.

The nuclear age left a frightening legacy. Dozens of those young women who painted radium salts onto the dials of clocks, tipping their brushes with their tongues, suffered cancers of the mouth and jaw. Marie Curie died of radiation poisoning. And the planet is poisoned by radioactive wastes from nuclear power plants and the making of bombs. On the upside, we now understand what makes the stars shine and why life on Earth is possible. "Let there be light," says the Bible. It is a nuclear light. Wonderful and terrible.

Face north. It is spring, and the **Big Dipper** is high overhead, pouring its refreshing waters on the Earth. No other group of stars is so easily recognized. The official name of the constellation is **Ursa Major**, the Big Bear. According to legend, the great god Zeus fell in love with the mortal Callisto, a huntress who roamed the wilds of Acadia in search of game. Hera, Zeus's wife, became jealous and turned Callisto into a bear. One day Callisto's son Arcas came upon his mother in ursine form and raised his bow to shoot what he took to be a bear. Zeus, looking down from Olympus, saw the impending tragedy and quickly changed Arcas into a little bear. He placed both mother and son safely in the sky where we see them today.

Hera had her revenge. She moved the Bears into the northern sky, where they revolve around the polestar, never setting, never resting. Follow the "**Pointer Stars**" at the front of the Dipper's bowl to **Polaris,** the one star that doesn't change position as the Earth turns. This is because Polaris lies almost exactly along the axis of the Earth; if you were at the Earth's North Pole, Polaris would be overhead. Many people expect Polaris to be the brightest star in the sky and are disappointed that it is not more prominent. The star is important only because of where it is: a steady marker for north. Use the handle of the Dipper as a way to find two other bright stars: "Make an *arc* to Arcturus; keep going and you'll *spy* Spica."

Look for the faint companion of the second star from the end of the Dipper's handle. Some people say this is the "missing" Pleiad, kidnapped from the Seven Sisters of the Pleiades by one of the Seven Brothers of the Big Dipper. The names of the pair are **Mizar** and **Alcor**. Also look for the **Three Leaps of the Gazelle,** paired hoofprints of faint stars "below" the Dipper's bowl.

The **Little Dipper—Ursa Minor,** the Little Bear—is an inconspicuous constellation. You will need a clear dark night to see all seven stars. Polaris, at the end of the handle, is the brightest.

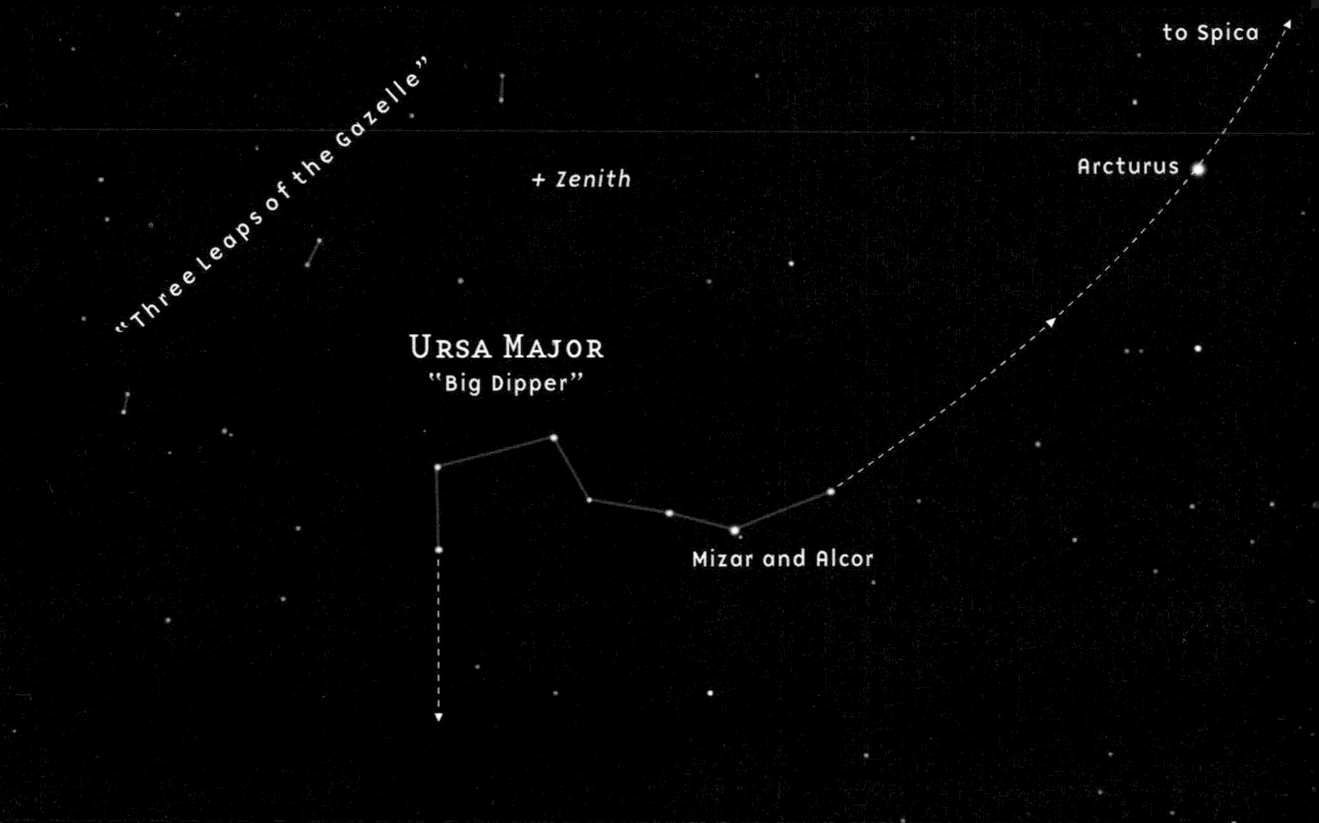

URSA MINOR
"Little Dipper"

Polaris

sky turns

LOOKING NORTH

It is a coincidence that we have a fairly bright star so close to the sky's (and above the Earth's) north pole. No star signposts the southern pole; mariners in southern seas have no steady beacon to guide them. Polaris has not always been a polestar, nor will it mark the pole forever. The Earth's axis has a slow wobble, like a top whose axis wobbles as it spins. The wobble (called **precession**) moves the poles around the sky in a 26,000-year cycle. Five thousand years ago the sky's north pole was near faint Thuban, in the constellation Draco. In 15,000 C.E. brilliant Vega will be near the sky's pole.

No need to worry about the wobble. The star maps in this book will be useful for your children and your children's children's children. And so on for many generations. But eventually the precession of the Earth's axis will put things askew. Also, stars have tiny motions of their own on the dome of night, called **proper motions** (motion proper to the star). Slowly, ever so slowly, proper motions distort the patterns of the constellations. One hundred thousand years from now, no one will see the shape of a water dipper in the stars of Ursa Major.

A few stars really zip across the sky. The telescopic star Groombridge 1830, near the first Leap of the Gazelle, is one of these "runaway stars" with atypically large proper motions. It streaks across the sky by a finger's breadth every 500 years on its breakneck journey through the Milky Way Galaxy.

Mizar and Alcor are not a true binary star system. They just happen to lie almost along the same line of sight. However, a modest-sized telescope reveals Mizar as a true binary, two white-hot stars of equal brightness held together by gravity. They are nearly 400 times farther apart than the Earth and Sun, and require many thousands of years to complete one revolution about each other. Mizar was the first star to be recognized as a binary system. In fact, the Mizar system is even more complex, as is shown by analysis of the star's light; what we see visually as one star, and telescopically as two, is actually a whirligig of five stars in complex orbits about one another.

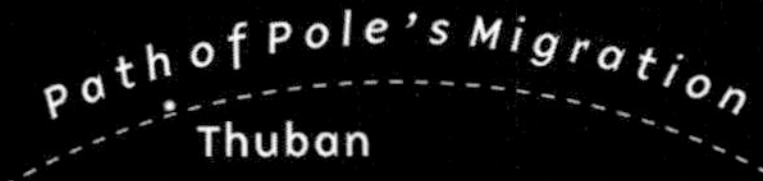

North Celestial Pole

Polaris

Vega

LOOKING NORTH

5. Spin

The Planets

One of Vincent van Gogh's paintings, *Road with Cypress and Star,* shows three celestial objects: a crescent Moon, a bright star, and a less bright star near the horizon. Art scholars agree that the canvas was painted at St.-Rémy, in southern France, near the end of van Gogh's yearlong stay at an asylum in that town. The artist left St.-Rémy on May 16, 1890, for Auvers, near Paris, where he committed suicide two months later. Using sky simulation computer software, we can work backward from the date of van Gogh's departure from St.-Rémy, looking for a likely arrangement of stars and Moon that served as a basis for his painting. (This was first done by astronomers Donald Olson and Russell Doescher, as reported in *Sky and Telescope* magazine in 1988; I repeated their simulation with my Starry Night Pro software from SPACE.com.) A crescent Moon occurred on April 19. The brilliant evening star Venus was near the Moon on that date, and little Mercury hunkered near the horizon. The arrangement of the three objects in the sky was strikingly similar to the objects in the painting, except that the order of

Jupiter's volcanic moon Io passes above the turbulent surface of the giant planet, casting a shadow on Jupiter's cloud tops.

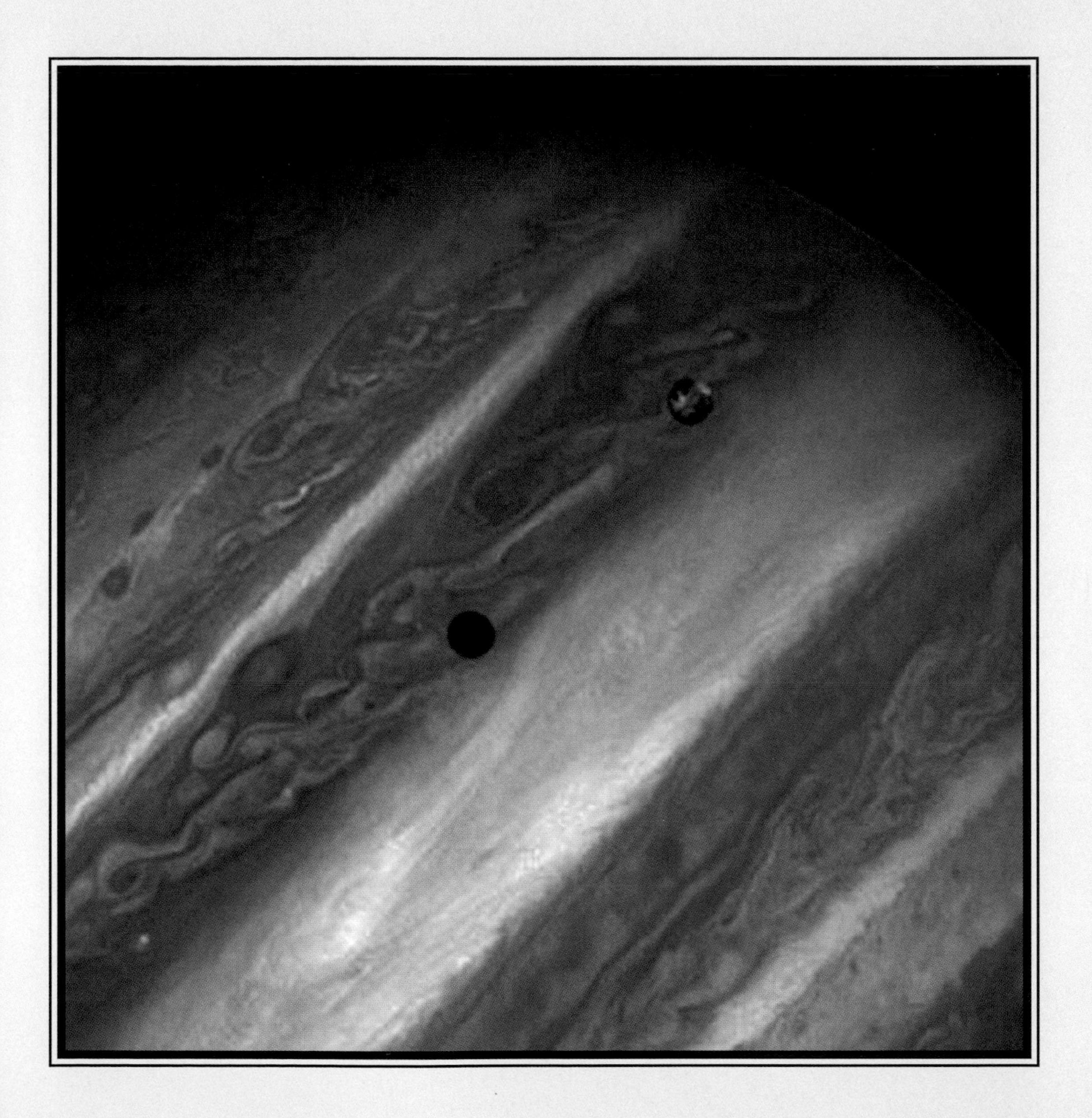

the objects is reversed and the Moon's crescent is tipped in a way that never happens in the real sky. Despite these differences, it seems likely that the April 19, 1890, conjunction of Venus, Mercury, and the Moon was the inspiration for van Gogh's painting.

A conjunction of the Moon with two planets would be of interest to any knowledgeable sky watcher. The Moon and planets are the movable feast of the heavens, shuttling back and forth on the loom of night. Five planets are visible to the unaided eye—Mercury, Venus, Mars, Jupiter, and Saturn—and since antiquity these "wandering stars" have teased our curiosity with their comings and goings. We always see them among the stars of the zodiac, the twelve constellations that define the Sun's apparent journey across the sky. Mercury and Venus stay near to the Sun; the other planets range widely. That the planets are guided in their ramblings by a grand design is obvious to even a casual observer. *Explaining* the pattern required the best efforts of astronomers from the Alexandrians to Kepler. We now know, of course, that the planets are not stars at all, but Earth-like objects in almost perfectly circular orbits about the Sun, shining by reflected sunlight.

Harvard astronomer Charles Whitney and UCLA art historian Albert Boime have considered other paintings of van Gogh's from the astronomical point of view, especially *Starry Night on the Rhone* and *Starry Night*. Whitney and Boime used planetariums to reconstruct past skies (they did their work before computers made it easy) and traveled to France to observe the sky from the places where van Gogh experienced it. They found several elements of scientific realism in the paintings. In *Starry Night on the Rhone* the Big Dipper is easily recognized, although the artist has placed the Dipper, a northern constellation, in a view to the southwest. Boime purports to find Venus and the constellation Aries among the stars of *Starry Night*. In the spiraling swirls of *Starry Night* both Whitney and Boime see the influence of astronomer William Parsons's 1845 drawing of the spiral nebula now known as the Whirlpool Galaxy. They guess that van Gogh may have seen a representation of Parsons's sketch in the works of the French astronomical popularizer Camille Flammarion. As Boime makes clear, van Gogh was keenly interested in astronomy, cartography, and science in general. He was also an exact observer of the night. In a letter to his sister, van Gogh says that "certain stars are citron-yellow, others have a pink glow, or a green, blue and forget-me-not brilliance"; stars do indeed exhibit these colors, but only to a careful observer. On this and other evidence, Whitney concludes that van Gogh had excellent night vision.

Nevertheless, van Gogh's skies are unlike any you or I have ever seen. His stars are whirling vortices of color, not cold points of distant light. Blue-black night yields in his paintings to torrents of yellow and green. Moons burn with the vitality of Suns. Space seethes with the energy of flame. Many people suppose that van Gogh's vertiginous nighttime paintings are a product of his madness. Art historian Ronald Pickvane rejects this interpretation: "Between his breakdowns at the asylum [van Gogh] had long periods of absolute lucidity, when he was completely master of himself and his art. That his mind was informed and imaginative, interpretive and highly analytical can be seen in the way he assessed his own work." Van Gogh's provocative visions of the night may not be products of madness, but neither are they literal representations of the sky. They represent (in Pickvane's words) "an exalted experience of reality."

We do not see in the night what van Gogh saw; we do not have his heightened sensitivity—his touch of madness. But every sky watcher knows the feeling of exhalation that comes with seeing a particularly beautiful gathering of planets. As I write this chapter, in the spring of 2000, a fine congregation of objects graces the evening sky. During the second week of April, red Mars zips along to pass Jupiter and Saturn, which are themselves drawing closer. At the beginning of the week, a thin crescent Moon adds its eyelash of light to the show, in a gathering reminiscent of van Gogh's *Road with Cypress and Star*, but with three planetary participants, not two. Other conjunctions follow: At the end of the month, Mercury passes Venus in the morning sky; in mid-May Venus kisses Jupiter in what might have been one of the most majestic pairings in decades were the planets not so close to the Sun in the morning sky; and in late May, Jupiter passes Saturn, but again frustratingly near to the Sun in our sky. I wait and watch, and when the time is right I take my students into the darkness to let those glittering objects draw us into their vortices. We are guided by science (my computer program shows us exactly what to expect), but what we hope to experience is something more than just two or three or four lights in the sky. Our experience embraces the planets and the twilight, birdsong and spring peepers, the scents of ripening spring, and the still night air—all five senses thrown open like windows to the world.

From the barred window of his room at the asylum at St.-Rémy, van Gogh had an unobstructed view of the night sky. His insomnia gave him ample opportunity to observe the stars. What he put on canvas was more than what he saw, and more than what a computer or planetarium can reconstruct. In one of his letters he wrote: "I should be desperate if my figures were correct. . . . my great longing is to make these incorrectnesses, these deviations, remodellings, changes of reality that they may become, yes, untruth if you like—but more true than literal truth." The painter Georges Braque said something similar: "Art is meant to disturb. Science reassures." The color-splashed, starry vortices of van Gogh's nighttime paintings certainly disturb. They disturb because they evoke something that in our less exalted way we recognize as truer than literal truth. The whirlwind stars of van Gogh's paintings draw us into a beautiful, terrifying, uncertain universe—a universe in which the individual must sometimes struggle to find security and meaning. Knowing that his wildly turbulent images contain an element of scientific realism is only mildly reassuring.

Almost every schoolroom has a solar system poster with the planets lined up like soldiers on parade, or a dangling mobile with the planets jostling each other in the breeze. These illustrations do little to help a student comprehend the true scale of the solar system or the emptiness of space. To get a better idea of our planetary system, imagine the Sun—our blazing star, nearly one million miles in diameter—as a smallish grapefruit on the goal line of a football field. Mercury is a fleck of sugar on the four-yard line, in nearly circular orbit about the Sun. Venus is a grain of salt on the seven-yard line, and Earth another salt grain on the ten-yard line. Mars is a sugar speck on the fifteen-yard line, Jupiter a pea on the fifty-yard line, and Saturn a slightly smaller pea at the far end of the field. (On this same scale, our nearest stellar neighbor, binary Alpha Centauri, is a pair of grapefruits one thousand miles away.) A second football field would carry us to peppercorn Uranus, and a third would reach peppercorn Neptune. It would take four football fields laid end to end to reach out to the average radius of mote-sized Pluto's highly eccentric orbit. Now, in your mind's eye, delete the ground, the stands, the goalposts, and everything else except the grapefruit and those nine tiny bits of food. Set the planets in motion about the Sun in that vast empty space of your imagination, and you have a pretty fair solar system. Sugar-fleck Mercury makes its journey around the grapefruit Sun in eighty-eight days; the Earth takes a year; and tiny Pluto, way out there nearly a quarter of a mile away, circles the Sun in a languorous 250 years.

Now imagine that NASA engineers on the salt-grain Earth, on the ten-yard line, decide to send a spacecraft to pea-sized Saturn, nearly a football field's length away. The mission will be called Cassini after the seventeenth-century astronomer who discovered the dark gap in Saturn's rings. In our football-field solar system, the spacecraft is infinitesimally small; in real life, it is about the size and weight of a school bus. It carries a probe to drop onto the surface of Titan, Saturn's largest moon. It is not possible with existing rocketry to make the craft move fast enough to travel directly to Saturn, at least not without taking decades to get there. So the engineers try a little orbital magic. Instead of directing the craft toward Saturn, they aim for Venus—in the opposite direction! The craft darts near Venus and gets a gravitational boost of energy at the expense of that planet, like a stone whirled in a sling. It coasts out past the orbit of Earth on an elongated orbit, then falls toward Venus again, where it gets another kick. By this time the spacecraft has made nearly two orbits of the Sun. Now it climbs again, away from the Sun, meeting Earth on the ten-yard line for another gravitational boost, then on to Jupiter out there on the fifty-yard line for one more increment of energy. Then it's on to Saturn.

If you can imagine the journey of the infinitesimally small spacecraft in the empty space of the football-field solar system—with salt-grain and pea-sized planets moving in circular orbits about the grapefruit Sun—then you have a notion of what a real space mission is like. Cassini was launched in October 1997. It made its first flyby of Venus in April 1998. It encountered Venus again in June of 1999, then whizzed past Earth in August of the same year. At the time of this writing, in the spring of 2000, the craft is on its way to a rendezvous with Jupiter in December 2000, and it will arrive at Saturn in 2004, more than six and a half years after launch. Of course, lots can go wrong on a flight such as this, and by the time you read these words Cassini may have another story to tell. But the important story here is the art and craft of navigation—the finesse with which navigational engineers take advantage of gravitational assists to nudge their spacecraft on their way, looping and darting like swallows through the vast emptiness of the solar system, hitching roundabout ways to their destinations. In the case of Cassini, the flybys of Venus and the Earth provided the equivalent of seventy-five tons of fuel. After the initial launch toward Venus, only small adjustments to the trajectory were necessary to direct the craft toward its next rendezvous.

This evening, Jupiter and Saturn are high overhead. I stand in their light and try to imagine that school-bus-sized machine that is sailing their way, one of an invisible flotilla of machines in space that represent humankind's attempt to know other worlds, making their circuitous ways to the objects of their investigation, stealing a gravitational kick where they can, stitching a wonderful embroidery into the fabric of the night.

Getting to Mercury is easier than going to Saturn. To visit Saturn, a spacecraft must climb "uphill" against the Sun's gravitational pull. But going to Mercury is "downhill" all the way, with the Sun's gravity pulling the spacecraft forward, and the problem is that when the spacecraft gets to Mercury it will be going too fast to slip into orbit about the planet. To date, there has been only one mission to Mercury—Mariner 10 in the mid-1970s. What Mariner 10 found when it got to Mercury was a hot, wasted planet with a densely cratered surface and only a wisp of atmosphere. Scientists still don't know much about the tiny red planet. Mercury orbits so close to the Sun that we never see it more than a few hand spans away from the Sun in Earth's sky. This means that Mercury is visible only in the twilight or the dawn, close to the horizon and therefore through lots of atmosphere—less than ideal conditions for eye or telescope. Even the Hubble Space Telescope is precluded from looking at Mercury, for fear that the Sun's brilliance might accidentally damage sensitive optical instruments. So little Moon-like Mercury skitters around the Sun on a short leash, never quite escaping the Sun's glare into the pitch black of night. No wonder the ancients named it for the fleet-footed god who passed back and forth between celestial and infernal realms.

Cloud-wrapped Venus, our nearest planetary neighbor, has been visited many times. The Soviets landed probes on the surface in the 1970s and 1980s, but the intense heat (hot enough to melt lead) and crushing atmosphere soon killed the electronics and silenced communication. Before several of the Soviet probes went dead, they sent back color pictures of a desolate rocky surface not unlike a terrestrial desert. In the early 1990s, the Magellan spacecraft mapped the hellish Venusian surface by radar from orbit above the clouds. It saw a desolate landscape threaded with lava rivers—what the Earth might have been if our planet had been closer to the Sun.

Mars has been the object of the most intensive exploration; of all the planets, it is most likely to support life—its own indigenous life, or someday human colonists. So far, space probes have found no little green men on Mars, nor even microscopic organisms, but "the game ain't over 'til it's over." Perhaps Mars has already been colonized by terrestrial microbes that hitched a ride on Martian landers, against the best efforts of space engineers to keep their spacecraft sterile.

The big, gassy outer planets with their numerous dancing moons have been visited, too. Jupiter's moons have been particular objects of curiosity, each with its own personality. Europa, certainly, will be the target of a future mission to search for life. A liquid water ocean may exist beneath Europa's frozen surface, and that ocean might have been entirely unfrozen early in Jupiter's history when the giant planet was hotter. "An ocean is the womb of a planet," says oceanographer John Delaney. If life exists elsewhere in the solar system, Europa may be the place to look. Saturn's large moon Titan will be another favored destination, with its thick atmosphere rich in nitrogen, methane, and hydrocarbons, and the Cassini spacecraft is blazing the trail. However, don't expect little green men on Titan; the surface is prohibitively cold.

The glorious pictures sent back from missions to the planets and their moons are in the backs of our minds as we watch the five naked-eye planets weaving among the stars. Ancient observers knew no difference between stars and planets, other than the fact that the latter objects move about. A half century of journeying across the great empty spaces has revealed a menagerie of magical worlds—like our own Earth in many ways, but also enough *unlike* the Earth to make us appreciate all the more the sweet benedictions of our blue-white planet.

All those planets. All those moons. Shepherded by gravity. Gravity: always tugging, gathering in. The whole universe would collapse into a heap if something were not resisting gravity—the mutual *attraction* of everything with mass. The explosive outward thrust of the big bang keeps the galaxies from drawing closer; if that outward momentum were ever expended, gravity would pull the galaxies back together. Fusion energy pushing out from the center of a star exquisitely balances the squeezing force of gravity, and keeps a star from collapsing on itself. And *rotation*—the reason that moons don't fall onto planets and planets don't fall into the Sun is rotation. Fill a bucket with water. Turn the bucket upside down over your head and you'll get wet. But swing the bucket round and round in a vertical circle over your head and the water stays in the bucket even when it's upside down. Rotation—call it *centrifugal force*—keeps the planets from falling into the Sun. Turn off the Sun's gravity, and the planets would fly off in straight lines, like pebbles from a sling. Stop the planets in their paths, and they fall into the Sun.

Our solar system has settled down into a steady balance of gravity and rotation. But it wasn't always so. Remember how a star is born. A vast cloud of dust and gas is compressed by gravity. As it shrinks, any rotational motion of the cloud increases (call it "the ice-skater effect"; when a skater pulls in his arms—brings his mass closer to the axis of rotation—he spins faster). Some of the mass of the collapsing nebula is spun off into a disk around the new star (call it "the pizza effect"; as the dough spins, it flattens out into a disk). This disk of whirling matter around a new star will become planets when gravity has done a bit more gathering together. We see disks of gas and dust around new stars in space. With the Hubble Space Telescope we can actually see planets aborning in the disks. Planets around stars seem to be the rule. But the shaping of a planet system is a chancy thing—a chaotic tug of war between gravity and rotation. And violent, too! Larger and larger chunks of matter in a protoplanetary disk are gathered by collision, until everything (or almost everything) finds its way into an object in a stable orbit—a planet, moon, or asteroid. The battered surfaces of Mercury and the Moon give us glimpses of the fierce bombardment these objects endured during the early days of the solar system. (Earth, too, was pummeled then, but atmospheric erosion and a mobile crust have erased all signs of that turbulent beginning.)

The violent times are not completely over yet. Four billion years after the major battering there are still rogue rocks flying around out there on looping paths that will sooner or later bring them into collision with a planet. One hundred tons of stuff from space fall onto the Earth every day, a steady rain of cosmic dust. What we see as "shooting stars" or "falling stars" are actually sand-grain-sized bits of rock or iron vaporized by friction as they streak into our atmosphere, just miles above the surface of the Earth. Every now and then the Earth gets knocked by something big enough to do real damage—witness the mile-wide hole in the ground near Winslow, Arizona, caused by a house-sized meteorite that smashed into the ground just 50,000 years ago—but these catastrophic impacts are rare enough that they do not prevent us from enjoying a good meteor shower or wishing on a "falling star." Planet watching is by and large an orderly activity. Planets move on utterly predictable paths, enhancing the beauty of the night with their ever-changing configurations. But streaking meteors, too, on any bright night can add their unpredictable thrill, reminding us what van Gogh saw so clearly: that chaos is the other face of creation.

Look to the east. Maybe wait a bit later in the evening or later in the season until you see bright Vega well above the horizon. Follow the *arc* of the Dipper's handle to find **Arcturus**, the third-brightest star in Earth's sky, and second-brightest star for most northern observers. (If you live as far south as Miami, Honolulu or Hong Kong, you can catch a glimpse of the second-brightest star, Canopus, below Orion near the southern horizon.) The name of the star means "Guardian of the Bear," an ancient connection to Ursa Major. Arcturus has an orangish hue; it is an orange giant star, more than twenty times more luminous than the Sun.

Some older people will remember Arcturus as the star that opened the Chicago world's fair in 1933. At that time Arcturus was thought to be forty light-years away, and another world's fair had been held in Chicago in 1893, just forty years earlier. Light from the star was focused on a photocell that activated a switch turning on the lights of the fair—light that had left Arcturus at the time of the previous exposition. Or so it was thought; more recent measurements show Arcturus to be about thirty-six light-years away.

Boötes *(boo-OH-teez)* is a kite-shaped constellation, but none of its other stars quite measure up to dazzling Arcturus.

Hercules is a large constellation of many stars; I have outlined only four, the so-called "Keystone." These are the easiest to find in the sky, and they are often taken to represent the torso of the Greek hero, offspring of Zeus and Alcmene in another of the great god's many liaisons with mortal women. On an exceptionally dark and clear night you might see a tiny blur between the two Keystone stars facing Boötes. Through a telescope, the blur explodes into a twinkling ball of 10,000 stars.

Between Boötes and Hercules is the constellation **Corona Borealis,** the Northern Crown, a semicircle of faint stars, with **Gemma** the brightest jewel in the crown. In Greek myth, this was the crown of Ariadne, who saved the Athenian hero Theseus from the Minotaur in the maze on Crete.

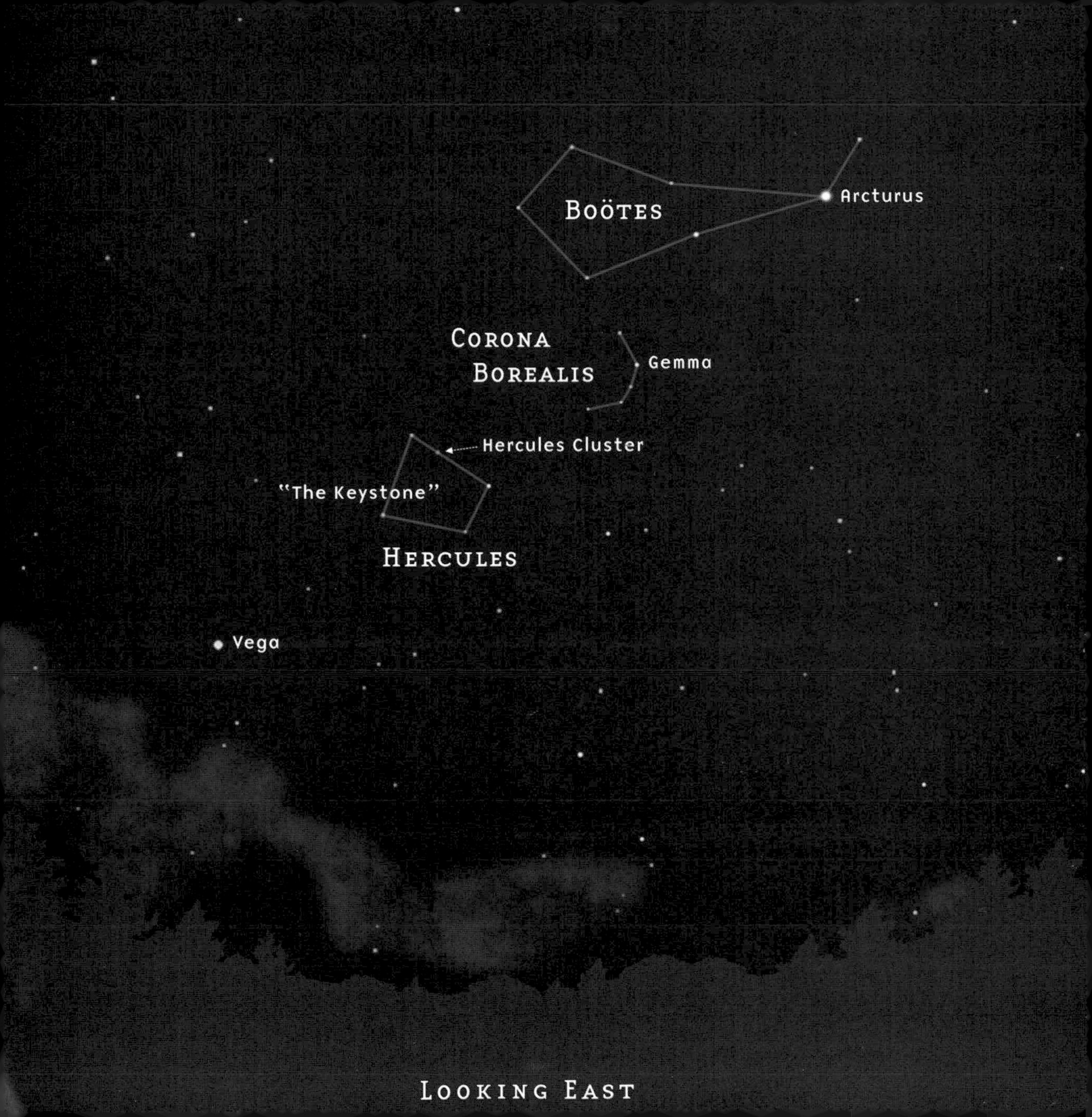

Boötes

Arcturus

Corona Borealis

Gemma

Hercules Cluster

"The Keystone"

Hercules

Vega

Looking East

As we saw in chapter 3 when we view nearby stars from different places in the Earth's orbit around the Sun, they seem to shift positions against the background of the more distant stars. This apparent shift is called parallax. The amount of shift tells us the distance to the stars. But the shifts are extremely small and can be observed only with sophisticated instruments. Arcturus dances back and forth by an amount 4,000 times smaller than the width of a common pin held at arm's length—and it is one of our nearer neighbors!

The fuzzy blur you might have seen in Hercules is a **globular cluster**, a ball-shaped buzz of 10,000 stars, held together by gravity and orbiting their common center of mass. About one hundred globular clusters hover above and below the plane of the Milky Way spiral. Actually, *hover* is not the right word; although on our time scale they seem to hang motionless in the sky, in fact they are in orbits about the center of the Galaxy. Other spiral galaxies also have halos of these mysterious clusters. If you live as far south as Miami, Honolulu or Hong Kong, you will have no trouble seeing the brightest of the globular clusters, Omega Centauri, a mysterious blur of light above the Southern Cross near the southern horizon. The resources section of this book will help you find it.

The globular cluster in Hercules is best known to professional and amateur astronomers as **M13**, the thirteenth object in a catalog of 110 fuzzy spots compiled by the seventeenth-century astronomer Charles Messier. Messier was a comet hunter, and comets appear as fuzzy spots in the sky. But comets move, even from night to night, and the objects cataloged by Messier stay put. He had no idea what they were. Today, we recognize the **Messier objects** as many kinds of things: galaxies, star clusters, remnants of exploded stars. Most of them require a telescope to see. Skilled amateur astronomers sometimes enjoy a dusk-to-dawn "Messier Marathon," trying to see all 110 objects in a single night, something that is possible only during a few nights in spring when the Sun is in a Messier-free part of the sky.

Arcturus

M13

Vega

LOOKING EAST

6. Swoosh

Comets and Meteors

A few years ago, I received an E-mail query from a young acquaintance. She wrote: "One of my favorite songs in all of explored space is 'Jupiter Crash' by the Cure. One of the lines is *'meanwhile millions of miles away in space/the incoming comet brushes Jupiter's face.'* Well [she continued], I just read an interview in which Robert Smith alludes to the Jupiter crash as if it were a real event. I had thought it was just something he had made up. So, if it is real, what exactly is a Jupiter crash?"

I had a pretty good idea what the "Jupiter crash" might be, but I wasn't at all sure about the Cure, Robert Smith, and the lyrics to his song. I turned to the Internet to do a little research. As probably everyone knows but me, the Cure is a British male rock band, formed in 1976, and still going strong in 1998 when I received my friend's query. Robert Smith was a founder of the group and remained at that time the band's central figure. The song "Jupiter Crash" is about a dramatic nighttime encounter with a woman on a beach, which the lyricist

THE TWENTY-ONE ICY FRAGMENTS OF COMET SHOEMAKER-LEVY 9, STRUNG OUT ACROSS 710,000 MILES OF SPACE, ON COURSE FOR A COLLISION WITH JUPITER.

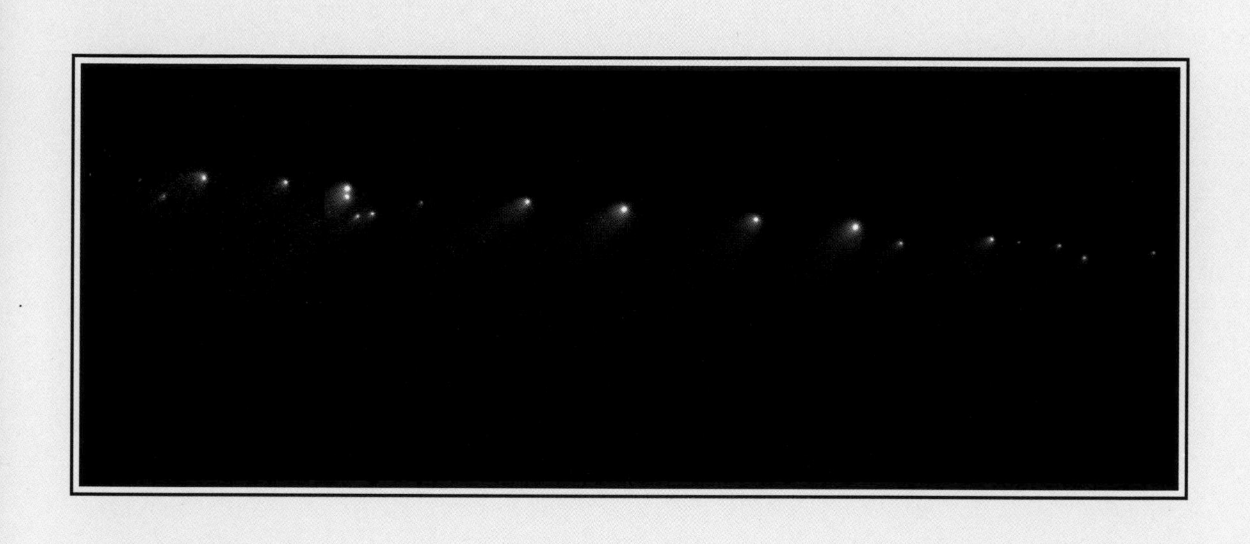

likens to a comet crash in space—an irresistible attraction, a violent splash into Jupiter's gassy sphere, the disappearance of the comet without a trace. "Is this how a star falls?" asks the lyricist. There is no doubt about it; Robert Smith took his metaphor from the Great Comet Crash of 1994, the most violent event ever witnessed in the solar system.

The comet that crashed into Jupiter was Shoemaker-Levy 9, discovered on March 25, 1993, by the prolific husband-and-wife comet-finding team of Gene and Carolyn Shoemaker, and ace comet hunter David Levy. The comet had recently undergone a close encounter with Jupiter. Astonishingly, the gravity of the giant planet had ripped the comet into a string of twenty-one pieces, strung out in space like beads on a string (see photo for this chapter). Nothing like it had ever been seen before. At discovery, the comet was still in the thrall of Jupiter's gravity. Astronomers quickly calculated that after a short excursion away from Jupiter, it would cycle back and crash into the planet in July 1994. Few astronomical events have excited so much anticipation among astronomers. The successive impacts of the chunks of Shoemaker-Levy 9, spread out over a week, were observed by countless telescopes on Earth and by the Hubble Space Telescope. The scars on Jupiter's face remained visible for almost a year.

Where do comets come from? Our solar system's orderly merry-go-round of planets is surrounded by a sphere of cold, dark comets that reaches halfway to the nearest stars. There are trillions of comets—snowballs of ice and dust, typically six miles in diameter, left over from the formation of the solar system.

Astronomers believe that these snowballs formed in the region of the giant outer planets 4.5 billion years ago and were driven into highy eccentric orbits by the gravitational influence of those planets (in much the same way as the Pioneer and Voyager spacecrafts were hurled out of the solar system by gravitational encounters with the giant planets). Town-sized chunks of rock and ice that might have become part of Jupiter, Saturn, Uranus, or Neptune were instead ejected to the cold nether regions far beyond Pluto. There they move with exquisite lassitude on lonely tracks, barely feeling the tug of the faraway Sun. Imagine the solar system in this new way: a star, burning with the fierce heat of thermonuclear fusion; nine spherical planets moving in more or less circular orbits; the whole thing enclosed in a thick, transparent shell of one trillion snowballs. In our football-field solar system of the previous chapter, with the Sun as a grapefruit on the goal line and the Earth as a grain of salt on the ten-yard line, it would take about 5,000 football fields laid end to end to reach the shell of comets. This comet-cocoon is called the Oort Cloud, after the Dutch astronomer Jan Oort who first proposed its existence in the 1950s.

Stars are not fixed in space. As the Galaxy turns, bearing all the stars with it, the stars move about within the Galaxy with more or less random motions (our Sun is drifting slowly in the direction of Vega). Every now and then—say every 100 million years or so—a star drifts close enough to the Sun to pass through the Oort Cloud. Some comets near the star's path are flung from the cloud into interstellar space, breaking their tenuous gravitational connection to the Sun; they are gone forever. Other comets are jostled from their orbits by the gravity of the passing star and sent showering inward toward the Sun. The passage of a star through or near the Oort Cloud can presage dangerous times on the planets, with more than the usual numbers of cometary collisions. Some of the many extinction events in Earth's history, recorded in the fossil record as a sudden disappearance of many species, may have been caused by a rain of comets perturbed from the Oort Cloud by a passing star. Even comets such as the bright recent visitors Hyakutake and Hale-Bopp may have been kicked from the Oort Cloud long ago by a passing star and nudged into their present orbits by close encounters with the giant planets. The next star that will penetrate the Oort Cloud is a small red dwarf called Gliese 710, which is scheduled to arrive in these parts 1.4 million years from now. It may send comets tumbling into the inner solar system; not a catastrophic rain of comets—the star is small—but certainly a sprinkle. To get some idea of what a minor comet collision might be like, consider the impact of the small comet that smashed into a remote region of Siberia in 1908, releasing the energy equivalent of a small atomic bomb.

The "Jupiter Crash" of July 1994 was the most violent planetary collision we know about: A snowball flung from the Oort Cloud by a passing star, snared by Jupiter, ripped into twenty-one pieces, then gobbled up piece by piece by Jupiter's gassy sphere. Each fragment of Comet Shoemaker-Levy 9 slammed into Jupiter at a speed of 140,000 miles per hour, raising a fireball that towered 2,000 miles above the planet, and leaving a huge, dark, Earth-sized cloud in its wake. "Yeah, that was it. That was the Jupiter crash. Drawn too close and gone in a flash," sings Robert Smith. The rock poets have not lost their contact with the grandeur of the sky.

Astronomers discover a few dozen comets annually with telescopes or on photographic plates. Only every few years or so does a comet grow bright enough in our sky to be seen with the naked eye. Hyakutake and Hale-Bopp were memorable comets of recent years, but they were not especially impressive compared to some comets of the past. The Great Comet of 1811 was visible in the sky to the naked eye for nine months, glowing for part of that time from dusk to dawn as brightly as the brightest stars. The Great March Comet of 1843 was briefly visible even in the daytime in proximity to the Sun. Later, that comet moved into the evening sky, as bright as Jupiter and with a tail that reached nearly halfway across the sky. Other magnificent comets appeared in 1858, 1861, 1882, 1910, 1927, and 1948. I remember with pleasure Comet Ikeya-Seki of 1965 and Comet West in 1976. I woke my children on the latter occasion to see the beautiful comet displayed against the light of dawn. Most memorable of all was the evening of April 3, 1996, when I gathered with a group of eager students in a broad dark field where we watched the Moon rise in full eclipse, a spooky pink pearl. Comet Hyakutake was in the northwest, showing a degree or two of tail, and Venus blazed near the Pleiades. Meteors added their unpredictable excitement.

Hyakutake was not as bright as other comets I have seen, but the combination of the comet, lunar eclipse, Venus in the Pleiades, shooting stars, and the company of enthusiastic young people made for a mix that I'll never forget.

Some comets have short enough periods (times for a complete cycle) to bring them around with steady regularity. Comet Encke has the shortest period of all—three and a half years—but it never reaches naked-eye visibility. Its elliptical orbit takes it from the vicinity of Mercury out nearly to Jupiter's orbit. The most famous short-period comet is Halley, which returns every seventy-six years. Its last visit in 1986 was best observed by viewers in the southern hemisphere. I watched Comet Halley under spectacularly clear skies at Ayers Rock in central Australia, and then again while floating on my back in a dark lagoon on Tahiti; there's something about comets that seems to call for special circumstances of observation. At Ayers Rock I was with a band of amateur and professional astronomers, armed with every kind of telescope and camera for taking a picture of the comet. They had set up their instruments at a dark spot in the desert and spent the night snapping away. I walked among them, occasionally taking a peek through their scopes, admiring their skill and enthusiasm. In the pitch dark the clicking of their shutters and the whirrings of telescope and camera-platform motors seemed like the sounds of desert insects. Their many dim red flashlights were like the eyes of nocturnal desert beasts. I wandered off on my own to a dark and sandy spot, where I stretched out on my back and faced the spangled canopy of stars. Halley glowed among them with a fuzzy light. It was not a spectacular comet, a mere slip of a thing, but in that desert context it was the apparition of a lifetime.

Although in 1986 Halley did not reach nearly the brightness it had achieved on previous visits, it was famously visited by five spacecraft, which obtained the first photographs of a comet's solid nucleus. What we saw was a spooky, coal-dark, potato-shaped object, nine miles long and five miles wide. This is what the Oort Cloud objects must look like when they are far from the Sun, where they lack their heat-induced comas and tails. The tail of a comet is its most characteristic (and variable) feature, a stream (sometime two streams) of volatile materials blown away from the comet by the radiation pressure of the Sun. Tail is not exactly the best word for this stream of particles; a comet's "tail" always points away from the Sun, so that when the comet is receding from the Sun, its tail actually runs before it.

Most of the comets that excite our admiration—such as Hyakutake and Hale-Bopp—are Oort Cloud comets, with orbital periods of thousands or millions of years. These comets follow long, cigar-shaped trajectories that take them far out beyond Pluto. They move like roller-coasters. At the cold, dark "tops" of their trajectories, far from the Sun, they proceed with a ponderous slowness. As they fall toward the inner solar system they gather speed, moving fastest as they zip around the Sun. Then they slow again as they climb back to the tops of their tracks. The appearance of Oort Cloud comets is unpredictable; they may have visited the inner solar system in the past, but so long ago that we have no record of it. A potentially bright Oort Cloud comet might be discovered at any time, typically somewhere near the orbit of Jupiter on its way to center stage in the inner solar system. The brightest apparitions occur when a comet passes near the Earth on its inward or outward journey.

There has never been a more lovely description of a comet's visit to the inner solar system than these lines of the poet Gerard Manley Hopkins, written on September 13, 1864:

— I am like a slip of comet,
Scarce worth discovery, in some corner seen
Bridging the slender difference of two stars,
Come out of space, or suddenly engender'd
By heady elements, for no man knows:
But when she sights the sun she grows and sizes
And spins her skirts out, while her central star
Shakes its cocooning mists; and so she comes
To fields of light; millions of traveling rays
Pierce her; she hangs upon the flame-cased sun
And sucks the light as full as Gideon's fleece:
But then her tether calls her; she falls off,

And as she dwindles shreds her smock of gold
Amidst the sistering planets, till she comes
To single Saturn, last and solitary;
And then goes out into the cavernous dark.
So I go out: my little sweet is done:
I have drawn heat from this contagious sun:
To not ungentle death now forth I run.

The poem was written as a speech in a play that Hopkins was writing called *Floris in Italy*, set in the Renaissance, at a time when Saturn was thought to be the outermost planet ("last and solitary"). The speaker, Giulia, is making her final farewell to Floris, who has brushed aside her attentions. She likens herself to a "slip of comet" who has fallen into the thrall of Floris's Sun, but now moves away "into the cavernous dark."

The astronomer David Levy, of Comet Shoemaker-Levy fame, wonders in the little book *More Things in Heaven and Earth* whether Hopkins based his poem on the observation of a real comet. Several comets appeared in the sky at about the time the poem was written. Donati's Comet of 1858 was one of the brightest comets of all time. Other notable comets appeared in 1861 and 1862 (Comet Swift-Tuttle, of which I will say more shortly). Levy chooses Comet Tempel, which appeared in the summer of 1864, as matching most closely the imagery of the poem. He guesses that it was during the predawn hours of August 4 that the young poet saw the comet while vacationing in Wales. On that day the tail of the comet stretched from the bright star Beta Tauri to the nearly as bright Iota Aurigae, "bridging the slender difference of two stars."

Wish upon a falling star and you are likely wishing upon a piece of comet. Round and round a comet goes in its orbit, and when it comes into the inner solar system the Sun's heat evaporates the dirty, dusty ices of its nucleus, causing a comet to grow a vast halo of glowing gases—"spins her skirts out," in Hopkin's lovely phrase—and a dusty tail that points into the distant darkness of its origin. All of this blown-away comet stuff continues in orbit about the Sun, but spread out before and aft of the comet, so that eventually after many orbits the trajectory of a comet become a dusty track. A typical comet loses about one-tenth of a percent of its mass every time it passes near the Sun. After one thousand passages or so, the comet might be completely vaporized, leaving behind only an orbiting stream of meteoric dust and pebbles. If the Earth in its annual journey around the Sun passes through one of these dusty tracks, we are treated to a *meteor shower.*

The most reliable of the annual meteor showers is the Perseids of late summer, peaking on or about August 11–12. The shower takes its name from the constellation—Perseus—from which the "shooting stars" seem to radiate. As we look toward Perseus, we are looking into the orbiting stream of dust that brings the meteors to us. What makes the Perseids special, in addition to their reliability, is their appearance on warm late-summer nights when we are often vacationing under dark skies. We spread our blankets on a beach or a mountain meadow and take in the pyrotechnics. At a typical Perseid peak we might see fifty or more meteors per hour. Each streak of light is made by a particle typically no larger than a grain of sand, heated to incandescence by friction and vaporized as it streaks into the Earth's atmosphere. The best time to view most meteor showers is after midnight, when we are on the side of the Earth that is plowing face-on into the meteor stream. The parent comet of the Perseids is Comet Swift-Tuttle, which last visited the inner solar system in 1992 (this was one of the comets that Hopkins might have seen in 1862); the years around 1992 were particularly rich in Perseids.

At least ten meteor showers occur each year that are worth watching for (see Appendix 2, "Meteor Showers"); the Geminids of mid-December and the Quadrantids of early January are especially likely to produce a worthwhile show. The Leonids of mid-November are generally not a particularly vibrant shower, but every thirty-three years or so they put on a spectacular display. The Leonid meteors are fragments of Comet Tempel-Tuttle (yes, Horace Tuttle was a prolific comet finder), which visits the inner solar system every thirty-three years on an orbit that takes it close to the orbit of the Earth. It is just after the comet has passed by that we are likely to experience a *meteor storm*, as Earth plunges through the stream of debris that trails the comet. During some historic Leonid storms the sky literally seemed to rain "stars." During Comet Tempel-Tuttle's

1966 visitation, observers in some parts of the world saw tens of thousands of meteors per hour, a once-in-a-lifetime feast of celestial fireworks.

The year 1999 provided thousands of meteors per hour for observers in Israel and Jordan. I watched all night with a large crowd of students from the observatory deck of our college. It was a cold night, but most of us were well wrapped up and supplied with hot coffee and cocoa. We didn't get the storm they experienced in the Middle East, but every few minutes a bright meteor streaked the sky, including occasional fireballs with glowing trains. With each streak, "Oohs!" and "Ahhs!" went up from the deck; people at other parts of the campus reported strange sounds all night from the direction of the science building.

But it was the previous year of Leonid watching that I liked best. Mid-November 1998 also promised the possibility of a meteor storm. Unfortunately, on the appointed night, the weather did not cooperate. The 11 P.M. weather report on television said there might be breaks in the clouds before dawn. That was enough to make myself and two of my students decide to camp out on the deck of the college observatory. We dragged out lounge chairs, wrapped up in blankets, and waited. It was a long, cold, damp night. Chris snored away as if he were home in bed. John and I passed in and out of sleep, tossing and turning in our discomfort, uncertain exactly why we were putting ourselves through this misery. We knew what we hoped for, but we also knew the odds were against us. The satellite photograph on the eleven o'clock report showed clouds backed up for hundreds of miles to the west. Even then, astronomers had predicted that observers in China would have the best view of the meteor storm—if they got anything at all. But we were game, and even when the drizzle began we pulled the blankets tighter around us and kept our eye on the western horizon, from where any clearance would come.

But it didn't clear, and when the first hint of dawn appeared in the east, we woke Chris, gathered our blankets, and made our ways home—without having seen a single Leonid. Nor did we hear one booming up there above the clouds.

In 1833, people not only saw a historic Leonid storm—up to 150,000 meteors per hour!—but also reported snapping, crackling, and popping noises, and occasional booms like cannon fire. According to astronomer Martin Beech, these were probably "electrophonic sounds," created by very-low-frequency radio waves generated by the interaction of a vaporizing meteoroid with the Earth's magnetic field. Only big and bright fireballs are likely to produce such noises, he says, objects bigger than three feet across and as bright as a nearly full Moon as they streak through our atmosphere just miles above the surface. Down below our clouds we heard nothing.

With or without sound, in storms or showers, meteors are of keen interest to astronomers: They bear clues to the origin of the solar system—and perhaps even to the origin of life (they are rich with carbon compounds). Tons of meteoric material from space smash into the Earth's atmosphere each day, mostly in the form of tiny particles that are vaporized by friction with the air. Larger objects can survive passage through the atmosphere and strike the ground. A meteorite on the ground is a scientific bonanza. In 1998, a group of European scientists reported finding meteor fragments incorporated in 1.4-billion-year-old sandstone from Finland, the oldest meteoric materials yet discovered on Earth, identifiable by their composition and heat-etched surfaces. These ancient particles of cosmic dust are gold mines of information about the history of the solar system during that long-ago epoch.

I recounted these tales of celestial noises and billion-year-old meteorites to John as we tossed and turned during our cloudy night on the observatory deck. It helped pass the time. The question remains: Why did we endure that night in cloud and drizzle? I'd give the same answer the naturalist Henry Beston gave when asked why he spent a year in a tiny shack on the Nauset Dunes of Cape Cod: *We sought a deeper sense that the creation is still going on*. "Creation is here and now," wrote Beston in *The Outermost House*. "So near is man to the creative pageant, so much a part is he of the endless and incredible experiment, that any glimpse he may have will be but the revelation of a moment, a solitary note heard in a symphony thundering through . . . time." What John, Chris, and I sought on the observatory deck was a glimpse of that incredible experiment, a few notes of the symphony of continuing creation. Up there above the clouds fragments of a comet made swan dives into the Earth's atmosphere, flaring briefly, scattering stardust, anointing the planet with the elements of life. We were disappointed that we didn't see—or hear—those bits of comet, but not so disappointed that we weren't ready to try again (with more success) the next year. Even cold and drizzle are part of the music of the here and now.

III. Summer

The Earth has now swung halfway around the Sun since we began looking at the winter sky. Again the Milky Way arches high overhead, and the second act of the celestial drama begins—brilliant stars and constellations afloat in a stream of galactic light. As we look toward Sagittarius and Scorpius, we are looking into the very heart of the Galaxy. Unfortunately, these constellations never rise very far above the southern horizon for most northern observers.

Scorpius, the Scorpion, is dominated by brilliant **Antares** *(an-TARE-eez)*. Because this star has a reddish hue, it was given a name that means "rival of Mars." It lies close to the ecliptic, so Mars does sometimes come this way; the red planet and the red star make quite a pair. I think of Antares as the red heart of the Scorpion. Look for the two claws, the body dipping close to the horizon, and **Shaula** *(SHAW-la)*, the Sting, at the tip of the curled tail. Native peoples of the South Pacific saw a fishhook dangling in the stream of the Milky Way. The Scorpion is the purported slayer of Orion, sent by the gods to punish the great hunter for his arrogant boast that he would kill all the animals on Earth. The gods then put Orion and the Scorpion among the stars, but on opposite sides of the sky so they would never fight again.

Sagittarius, the Archer, has no very bright stars, and like Scorpius it is hard to see unless you have a clear southern horizon. However, as a zodiac constellation it is well worth knowing. The Archer is a centaur—with the head and torso of a man and the body of a horse—one of two centaurs in the sky. (The other is Centaurus in the southern sky.) In Greek mythology, Sagittarius is immortal Chiron, the mildest and wisest of the centaurs. In a tragic accident, he was shot with a poisoned arrow by his favorite pupil, Hercules. To escape the excruciating pain, he renounced his immortality and accepted death. Zeus then placed him among the stars. Don't look for the figure of a centaur; you will never find it. Look instead for a **Teapot**, pouring its contents onto the Scorpion's tail.

A third zodiac constellation is in this part of the sky—inconspicuous **Libra**, the Scales, the only inanimate figure in the animate zoo of the zodiac. There are eight and a half non-human animals and four and a half humans in the zodiac. Can you guess them all?

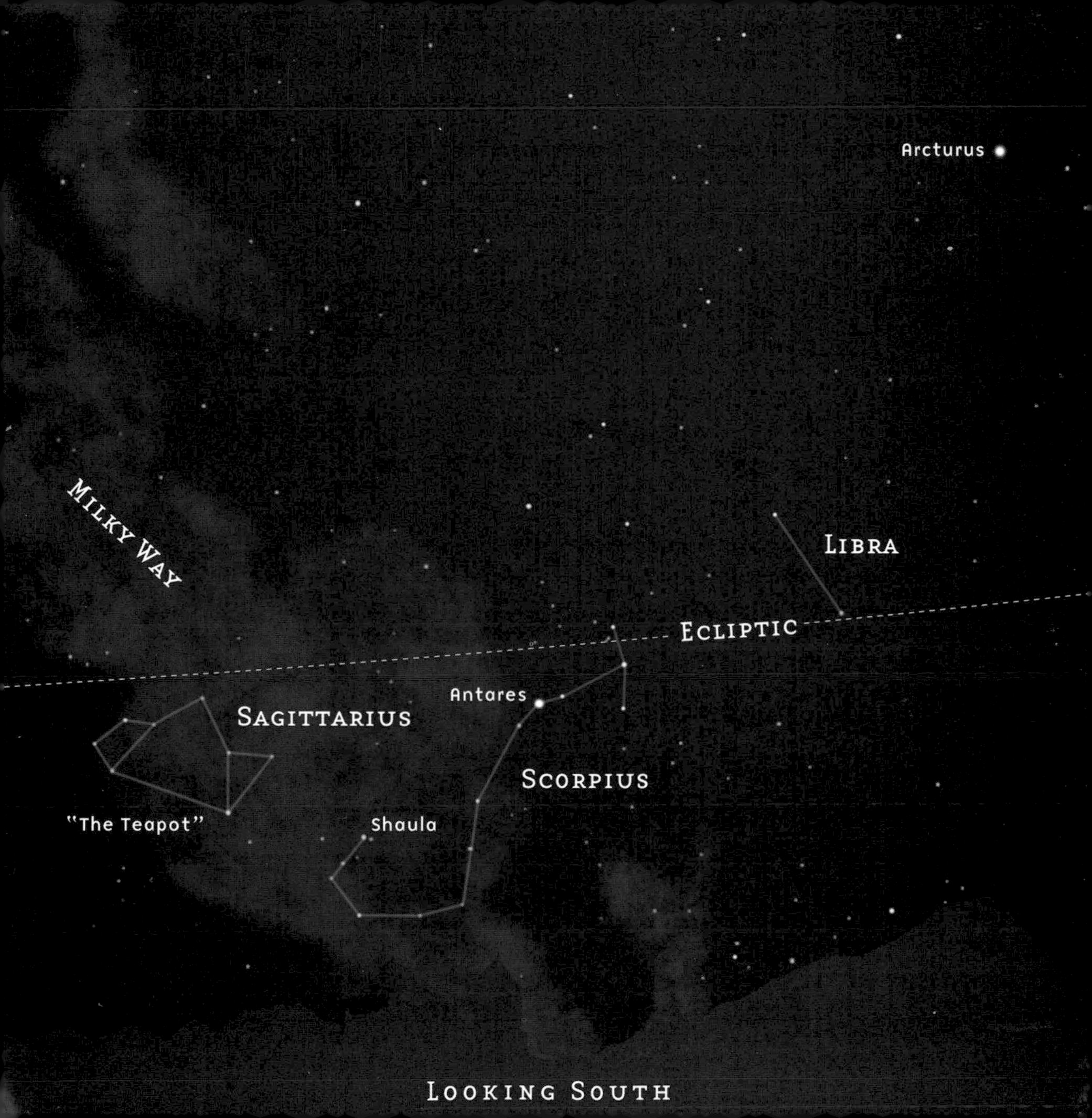

Arcturus

Milky Way

Libra

Ecliptic

Antares

Sagittarius

Scorpius

"The Teapot"

Shaula

Looking South

Close to the Teapot's spout lies the hidden center of the Milky Way Galaxy, 30,000 light-years away. This region of the sky is a telescopic paradise, dense with star clouds and clusters, glowing star-birthing nebulae, and dark banks of dust and gas. More Messier objects can be found just north of the Teapot than in any other part of the sky—a veritable mist of "fuzzy spots" rising like steam from the Teapot's spout.

At the very center of the galactic whirlpool lies an intense source of radio, X-ray, and gamma-ray energy known as **Sagittarius A**. Most astronomers believe this is associated with a gigantic black hole with a mass of a million Suns. A **black hole** is a dense, compact object with a superpowerful gravitational field. Matter can be drawn into a black hole, and as it accelerates into the maw it radiates intense energy. But nothing, not even light, can escape. Apparently, stars, dust, and gas were pulled together early in the universe's history to form this monster at the core of our galaxy—-and other galaxies too. As these galactic black holes formed, they emitted the fierce radiation we see today as the distant (and therefore ancient) **quasars.**

Antares is one of the largest stars in the sky, a bloated red giant at the end of its life, rivaled in size only by Betelgeuse in Orion. If Antares were where the Sun is, its surface would extend nearly to the orbit of Jupiter! However, the entire body of the star is not distended. The core has collapsed to become a supercompact, blazing-hot nuclear furnace where hydrogen and helium are fused into heavy elements such as carbon, nitrogen, oxygen, and iron. If the star dies convulsively as a supernova, it will eject these elements into space, to become part of interstellar nebulae out of which future stars and planets will be born. Every heavy-element atom in the body of the Earth—and in *your* body too—was forged in a supergiant star and blasted into space long before the Sun and its planets were born.

The two brightest stars of Libra are called the Northern Claw and the Southern Claw, and in ancient times they were part of Scorpius. The Romans cut the claws off the Scorpion and made a new constellation representing the Scales of Justice. The Greek astronomer Eratosthenes supposedly listed the faint Northern Claw as brighter than Antares. A mystery! Has one of these stars changed brightness since ancient times?

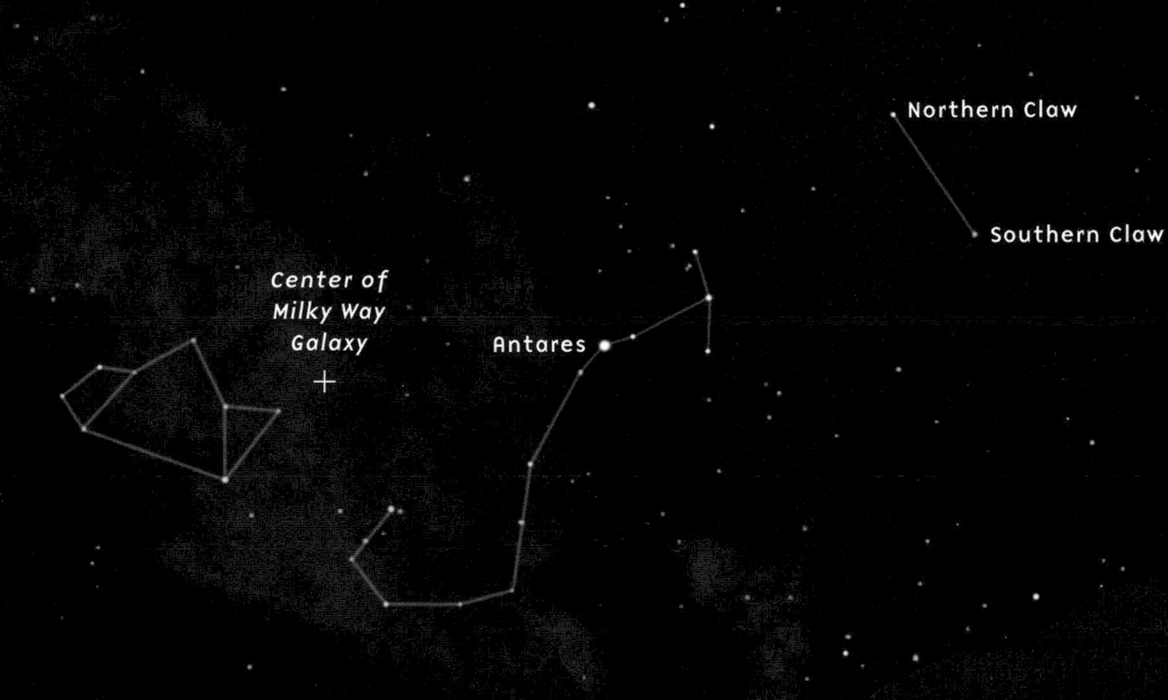

LOOKING SOUTH

7. Shadow

The Moon and Eclipses

At 6:27 P.M. this evening, June 16, 2000, the Moon will be full. If the clouds stay away, it will rise in the east at sunset, huge, golden, glorious. The full Moon of June is called the Rose Moon, Flower Moon, or Strawberry Moon. It might also be called the Honey Moon, since June is the traditional month of marriage. The "honeymoon" was originally the first month after the wedding, when (according to Samuel Johnson) "there is nothing but tenderness and pleasure." Renaissance writers often pointed out that the honeymoon, like the Moon itself, is no sooner full than it begins to wane—June brides and grooms beware! (On the morning of my own wedding, in 1958, an almost new Moon rose with Venus, the goddess of love, just before the Sun. If I had a Renaissance frame of mind, I would have taken this as a propitious sign; a new Moon can only grow fuller.) Guy Ottewell, who publishes an annual calendar of celestial events (see "Resources"), believes that the word for the postnuptial month (and today's wedding trip) might have been suggested by the color of the midsummer full Moon: honey colored because it is seen lower in the sky. In late June the Sun is as high in our northern skies as it ever gets, and the full Moon, which is always directly

Astronaut Edwin Aldrin walks on the Moon near a leg of the Lunar Module during the Apollo 11 mission.

opposite the Sun, takes a low path across the dome of night. Even at midnight it stands not far above the southern horizon, tinged golden by the atmosphere. Newlyweds celebrate its honey light.

The time is mostly past when the light of a full Moon usefully extended the activities of day, but sky lore still reflects the old ways. The full Moon of July is called the Hay Moon; its light gave farmers extra time to bring in the hay before rain. The full Moon of May is the Planting Moon, and the full Moon nearest the autumn equinox is the Harvest Moon. The next full Moon after the Harvest Moon is the Hunter's Moon; the fields have been reaped and hunters can more easily see their prey. Two hundred years ago, before electricity, a full Moon made it possible to travel safely at night. In England, a group of entrepreneurs, including Josiah Wedgwood (of pottery fame), Matthew Boulton (of steam engine fame), and Erasmus Darwin (of grandson fame), established a "Lunar Society" that met each month on the night of the full Moon to socialize and exchange ideas. In the course of their moonlit meetings they helped create the Industrial Revolution and launched humankind upon a course of middle-class democracy and artificial light.

But who looks anymore at those little symbols of the Moon's phases that decorate our calendars? Our year is in sync with the Earth's orbit around the Sun, and our clocks follow the daily spin of the Earth on its axis, but the Moon symbols wander all over the pages of the calendar. Originally, calendar months began with new Moons—as they still do, for example, in the Islamic calendar—but the correlation of month and Moon was lost in the Latin world when Julius Caesar promulgated his calendrical reforms in 45 B.C.E. A new Moon can now occur on any day of the month, and the day varies from year to year. And so we have lost almost all of our ancient cultural connections with the Moon. No one makes hay by moonlight anymore because no one makes hay; huge machines bundle the mown grass, rain or shine, in plastic wrap for silage. Menstruation may occur monthly in response to some ancient entrainment by the Moon, but no woman watches the Moon to predict her flow. Only sailors who live by the flow of tides still mind the Moon in its phases.

Or rather sailors and *noctivagants* (a wonderful word meaning "nighttime walkers"). Each summer month when the Moon is full, I take to the woods and meadows. It's the time of the big-eyed creatures: woodchucks, opossums, raccoons, foxes, skunks, and owls. It's the time of the music makers: bullfrogs, crickets, and amorous woodcocks. It's bat time. Glowworm time. Snail and slug time. The world is different by moonlight—all flutter, scuttle, and slink. Each turn in the path is like a card turned over in a game of showdown; you don't know what you will find there—a rush of wings, a pair of eyes. Henry David Thoreau was a nighttime walker. The outdoors at midnight, he said, is as unknown to most of us as Central Africa. He didn't waste full Moons. "What if one moon has come and gone with its world of poetry, its weird teachings, its oracular suggestions, and I have not used her?" he wrote. Walking by moonlight, Thoreau felt a tide in his thoughts. A Moon tide. Pulling him upward toward the inverted bowl of night. "How insupportable would be our days," he wrote in an essay on night and moonlight, "if the night with its dews and darkness did not come to restore the drooping world."

No other inner planet has a moon as big and bright as ours. The tiny Martian moons, Phobos and Deimos, provide too little light for midnight walks on that planet. Phobos sheds about as much light on the surface of Mars as Venus does on Earth. The light of Deimos is comparable to the light of a star. Earth's Moon is improbably big. Mysteriously big. The full Moon sheds 1,000 more light on the surface of the Earth than all the stars together. It is 400,000 times less bright than the Sun, but it provides adequate illumination for nighttime walking. And that's why noctivagants want to know when the Moon is full, especially on warm summer nights. Like Bottom in Shakespeare's *Midsummer Night's Dream,* they pay attention to those little lunar symbols on the calendar: "A calendar, a calendar! look in the almanac; find out moonshine, find out moonshine."

The Moon's size compared to the size of Earth is greater than that of any other satellite to its planet in the solar system, with the exception of Pluto and its mysterious companion Charon. A full Moon is fine for honeymooners and noctivagants, but for astronomers, professional and amateur, the Moon's light can be a nuisance, especially when they go looking for comets, meteors, and other faint lights that sometimes grace the night sky. And if the big terrestrial Moon is a bother to sky watchers, it is a positive embarrassment to the theoretical astronomers who try to tell us where it came from. Theories for the origin of the Moon have generally stumbled upon the block of the Moon's outlandish size.

In the past, three kinds of theories have been evoked for explaining the Moon's origin. They can be classified by calling the Moon the "sibling," the "child," or the "spouse" of Earth. The sibling theory assumes that the Earth and the Moon condensed together from an eddy in the larger whirlpool of accreting materials that became the solar system. The child theory assumes that the Moon's material was "spun off" from the outer layers of a rapidly spinning Earth, early in the Earth's history before it solidified. The spouse theory assumes that the Moon formed somewhere else in the solar system and was subsequently captured by the Earth's gravity. Each of these theories has dynamic problems that are not easily resolved. Recently, a fourth theory of the Moon's origin has gained strong support. This theory assumes that early in its history the Earth suffered a grazing impact by a Mars-sized object. The collision blasted into Earth orbit a mass of molten materials, partly from the Earth, partly from the colliding object, which subsequently solidified to become the Moon.

The impact theory agrees with current ideas about the formation of the solar system, which apparently began as a whirlpool of gas and dust around a new star and condensed in stages. First, the gas and dust collected gravitationally into pea-sized objects. Then the "peas" gathered into chunks the size of buildings. The "buildings" collided to make bigger bodies, and so on until the present planets came into being. In the last stages of this process, a few very large impacts are to be expected. One of those massive terminal impacts may have splashed the Moon into being. A primary scientific objective for the Apollo missions to the Moon was to discover the secret of the Moon's origin. The missions made it clear that there are intriguing similarities and differences in the chemical composition of the Earth and Moon, and revealed some clues to the Moon's internal structure. The missions refined the parameters within which a successful theory for the Moon's origin must be found, but they did not definitively resolve the riddle. Nevertheless, the "big splash" theory explains why the Moon's density is similar to the outer layers of the Earth, and why there are subtle differences in composition between Earth and Moon. It also accounts for the untypical size of Earth's satellite.

Recent progress toward learning about the Moon's origin has mostly come from using the laws of physics to model the origin of the Moon on computers. I've seen screen shots of these computer simulations of the titanic collision that may have given us our Moon. On the computer screen, the early Earth and the impacting Mars-sized body are color-coded for clarity, with metallic cores and rocky mantles further distinguished by color. The impactor comes hurtling in at 18,000 miles per hour. At time $t = 0$ they touch, with the impactor hitting off center. At t = 400 seconds, the impacting body has smushed into the Earth's mantle, and both objects have begun to melt from the kinetic energy released by the collision. At t = 800 seconds, vaporized rock from both Earth and impactor begins to squirt outward. Thirty minutes after impact, the two bodies have mixed their molten materials, all the way down to the Earth's core, and a stream of vaporized rock from the Earth and impactor begins to splash away from Earth. This stuff goes into orbit around the Earth and coalesces to become the Moon. The Moon, being smaller, solidifies more quickly than the Earth. But before Earth and Moon solidify, gravity pulls them into spherical shapes. Watching all of this happen on the computer screen is almost like being there 4 billion years ago when the Earth was clobbered. It is possible that the impact tipped the Earth's axis of rotation. If so, our gentle seasons had their origin in violence.

And so the Earth acquired its Moon—its Venus to the Sun's Apollo—a silver metronomic disk that in its monthly cycle sets the tempo of the tides, the comings and goings of sea creatures, and patterns of human activity before the coming of electric lights. By our modern accounting, it is just a coincidence that the Sun and Moon appear to be the same size in the sky; the Sun is 400 times bigger than the Moon and 400 times farther away. For our ancestors, nothing was "just a coincidence." For them, the Sun and Moon were matching orbs that defined the cycles of their lives—divine presences, the one hot and constant, the other cool and variable. And no events were more fraught with portent than eclipses of the Moon or Sun, those terrifying times when one or the other of the godlike presences hid its face. In our more scientific age, we know that eclipses are the inevitable consequences of one object moving in front of another, a matter of shadows, precisely predictable. We no longer cower in trepidation when the Moon or Sun goes dark; instead we are freed from superstitious fear to exalt in an eclipse's uncommon beauty.

Another food analogy. This time, think of the *Earth* as the grapefruit. On the same scale, the Moon would be a grape about twenty feet away. Both Earth and Moon cast shadows, long, conical wizard's caps of darkness that point away from the Sun. (Shelley's "pyramid of night/ which points into the heaven.") The Earth's cone of night reaches out about three times farther than the Moon. The grapefruit Earth's cap of darkness fits snugly about its brow and extends to an apex sixty feet away—a tall, skinny wizard's cap, indeed! Now here's the important thing: The plane of the Moon's orbit is slightly tipped with respect to the plane of the Earth's orbit around the Sun (the so-called plane of the ecliptic, from eclipses)—tipped like a wobbly wheel. Most months, when the Moon circles behind the Earth, it passes just above or just below the Earth's long, skinny shadow. Only twice a year or so, when the Moon is near its passage through the ecliptic plane, does it slip through the Earth's shadow—the grape through the grapefruit's cone of darkness. For an hour or so the Moon's face, which was full and bright, goes dark. Anyone who resides on the night side of the Earth can see the eclipse.

What is most striking about a total lunar eclipse is the color of the Moon, not completely black, but tinged a deep red or copper by sunlight that is bent into the Earth's shadow by refraction through the Earth's atmosphere. The grapefruit Earth has a tissue-thin wrap of air, and a few rays of reddish light wend their way through the air around the curve of the Earth onto the face of the Moon in shadow. The color of this refracted light cannot be exactly predicted—it depends on how much cloud, volcanic dust, and pollution are in the atmosphere—so every total lunar eclipse has the possibility of surprise. Total eclipses of the Moon are not particularly rare. Everybody on Earth has a chance to see one every few years or so—if they take the time to look. In January 2000, I watched a total lunar eclipse from the island of Exuma in the Bahamas, on the tropic of Cancer. The vantage point couldn't have been better. At mideclipse, the island was almost exactly at the center of the Earth's night side, front row center, with the Moon close to our zenith. I lay on my back in the warm tropic air and watched the Moon's disappearing act unfold as if it were doing its thing for me alone.

Total eclipses of the Sun are something else again. No natural phenomenon is more breathtaking, more filled with wonder. And rare! The average time you will have to wait at any given place on Earth for a total solar eclipse is 375 years!

Partial solar eclipses—when the Sun's disk is partly obscured—don't count. The difference between a total solar eclipse and a 99 percent partial solar eclipse is the difference between night and day. In former times, most people lived out their lives without witnessing this most extraordinary of celestial events. In our day of convenient and relatively inexpensive travel, no one should miss the chance to see a total solar eclipse at least once.

On August 11, 1999, I was with 800 other avid eclipse chasers, including a high-tech team from NASA, on a cruise ship in the Black Sea, waiting for a total eclipse of the Sun. Our location offered several advantages: First, it was close to the place of maximum totality (two and a half minutes); second, it had a better than average chance of clear skies; and third, the ship's mobility would make it possible to seek a hole in the clouds in case of inclement weather. The Moon's shadow first touched Earth in the Atlantic Ocean south of Newfoundland, then swept eastward across the southwestern tip of England, central Europe, the Black Sea, Turkey, Iran, Pakistan and India—a racing dot of darkness on the face of the globe. The lucky people of Stuttgart, Munich, and Bucharest could sit in their back gardens and watch this eclipse—if skies were clear, which they weren't for most of Europe—but the rest of us had to make a journey to a faraway place.

Most people know why a solar eclipse happens: The Moon blocks the Sun's light. What is less well understood is why total solar eclipses are so rare. Remember the grapefruit Earth with the Moon as a grape twenty feet away? Like the Earth, the Moon wears a wizard's cap of darkness—lunar night. To have a total eclipse of the Sun, the Moon's conical shadow must touch some part of the Earth's surface. And here's the kicker: The Moon's shadow is almost exactly as long as the average distance of the Moon from the Earth (think of the grape with its twenty-foot-long shadow tapering to a point near the grapefruit Earth). Because of the tilt of the Moon's orbit with respect to the plane of the ecliptic, most months the tip of the Moon's shadow passes above or below the Earth, but about twice a year, when the Moon is near the ecliptic plane, a solar eclipse is possible. But it's not as sure a thing as an eclipse of the Moon. The Moon moves around the Earth in an almost perfectly circular orbit, but the Earth is not quite at the center of the circle. Sometimes the Moon is a bit farther from the Earth, and sometimes a bit closer. When the Moon is near apogee—its greatest distance from the Earth—the tip of its shadow does not quite reach to the Earth's surface and a total eclipse cannot occur. When the Moon is at perigee—its nearest distance to Earth—the rapierlike tip of its shadow just reaches Earth or even extends a bit beyond. To see a total solar eclipse, one must be in the narrow path where the rapier's tip slices the surface of the Earth.

If the Moon were a bit smaller or a bit farther away, we wouldn't have solar eclipses at all. And if the Moon were bigger or closer, the intersection of the shadow with Earth would be larger and eclipses wouldn't be so rare. By celestial coincidence, the relative sizes and distances of the Sun and Moon are such that we are graced with an extraordinary event that is deliciously rare. Europe will not have another total solar eclipse until 2027, and then only Gibraltar will be touched. India and China will have an eclipse in 2009, but Japan must wait till 2035. The United States will be blessed in 2017, when the Moon's shadow cuts right across the heartland from west to east. Then North America will have to wait until 2024, when the Moon's shadow will sweep up from Mexico, across the United States, and into Canada.

Take a twelve-inch diameter terrestrial globe such as you might have in your home or schoolroom, and every year or so draw a random line ten or twelve inches long across its face with a black felt-tip marker. The line can be anywhere from north pole to south pole and in any hemisphere. These marks are typical of the paths of total solar eclipses. How long until the entire globe is painted with shadows? That is: What is the *longest* time that any place on the Earth's surface would have to wait for a total solar eclipse? Mathematical astronomer Jean Meeus has done the calculation, and the answer turns out to be 4,500 years. Month by month, year by year, the Moon goes round and round the Earth and now and then brushes the surface of the Earth with the featherlike tip of its shadow. Along that line, the Sun goes dark, stars come out, the air cools, and birds sing, thinking it is twilight. And nothing, nothing you can see in the heavens in your lifetime will be more exciting than seeing the face of the Sun become an inkpot hole in the sky.

That's why I was on that ship in the Black Sea with 800 other people. We were waiting in a calm sea under cloudless skies when the dot of darkness raced out of Europe and across the water, extinguishing the Sun's light for two and a half minutes. With a last blaze of glory—like the gem of a diamond ring—the

Sun's disk became jet-black. Streaks of radiance streamed outward into a blue-black sky, and crimson flecks marked solar storms leaping beyond the rim of the covering Moon. Venus blazed nearby. The horizon all around was rosy with a midday dawn. When the Sun went dark, 801 jaws dropped and eyes gaped wide. The Second Coming could hardly have evoked more awe.

The summer Milky Way rises in the east like a luminous screen. You will not see it in city lights, nor if a bright Moon is in the sky, but everyone should at some time in their life make their way to a place where the night sky is ultradark and admire the river of light that was once so prominent a part of our ancestors's experience—in the days before artificial light and smog obscured the stars. Even in the city, however, you will see the three stars of the "Summer Triangle": Vega in Lyra (the Lyre), Deneb in Cygnus (the Swan), and Altair in Aquila (the Eagle). These stars will dominate our evening skies right through late summer and autumn.

Blue-white **Vega** *(VEE-ga* or *VAY-ga)* takes its name from an Arabic word for "swooping" and suggests that before this group of stars became a Greek lyre it was the third bird in the summer sky. If you live near midlatitudes in the northern hemisphere, Vega will pass almost directly overhead, like its winter counterpart Capella. It is the only bright star in **Lyra**, an otherwise inconspicuous constellation. Look for a little parallelogram of stars to the south of Vega.

Cygnus, the Swan, flys south along the stream of the Milky Way. Look for its long neck aiming between Lyra and Aquila, outstretched wings, and short tail. **Deneb** *(DEN-eb)* is the Arabic word for "tail." (You will remember Denebola at the tail of the Lion.) Many people look for the shape of an old-fashioned kite, but then Deneb is at the *top* of the kite, not the tail. Sometimes Cygnus is called the Northern Cross, a symbolism that is most apt in early winter when the cross stands almost vertical on the northwestern horizon. The Milky Way near Cygnus is rewarding to scan with binoculars.

Aquila *(ACK-will-uh),* the Eagle, flys north along the Milky Way; if we could set the two great birds in motion, they would pass in the night. In Greek myth, the Eagle was the faithful messenger of Zeus. **Altair** *(al-TARE)* means "the flying one."

If you want to count yourself a connoisseur of constellations, learn to recognize **Sagitta** *(sa-GIT-ta),* the Arrow, and **Delphinus** *(del-FINE-us),* the Dolphin. These faint but engaging figures dart before the Eagle, one to either side.

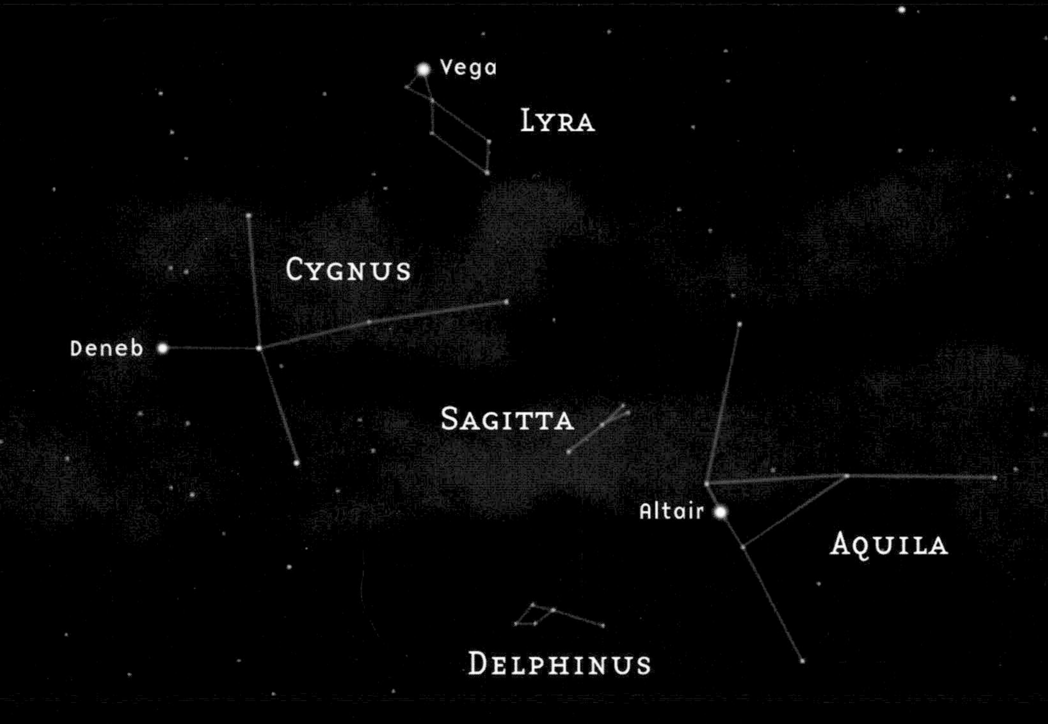

LOOKING EAST

As Vega passes overhead on a late summer evening, Sirius is almost directly below your feet. The two stars are similar blue-white giants. Sirius appears brighter than Vega because it is closer to Earth. Of the stars we have named so far, Vega is the fourth closest—twenty-seven light-years—after Sirius, Procyon, and Altair. Hundreds of tiny, faint red dwarf stars lie closer than Vega, invisible to the unaided eye. Red dwarfs are the most populous kind of star in the Galaxy, but the less numerous blue and red giants dominate our sky. Most of the stars we see with the unaided eye are more intrinsically luminous than the Sun—stars near the top of a pyramid of brightness.

The lovely **Ring Nebula** (M57 in Messier's catalog) lies between the two stars at the southern side of the Lyra parallelogram. A good amateur telescope shows a smoke ring adrift in the dark. The ring is actually a bubble of gas blown off a star during the convulsions that accompany the end of a big star's life. At the center of the ring, visible on observatory photographs, is the white-hot core of the star that expelled its outer layers.

You will notice on the sky map a gap in the Milky Way that lies more or less parallel to the body of the Swan. This **Great Rift** is caused by a thick band of dust and gas that blocks the light of more distant Milky Way stars—part of the Galaxy's store of dust and gas out of which new stars are born, and to which stars return some of their substance when they die violently. Along the Swan's neck is a "cloud" of new stars, the Cygnus Star Cloud, a stunning sight through binoculars.

One weekend of August 1975, a bright new star suddenly appeared near Deneb in Cygnus, almost as bright as Deneb itself, like a feather falling from the Swan's tail. The star quickly began to fade; within a week it was no longer visible to the naked eye. A dying star, thousands of light-years across the Cygnus arm of the Galaxy, had suddenly flared as a **nova**, puffing off a bubble of its outer layers. To date, this was the last bright naked-eye nova in northern skies. Another could occur at any time—one of the great, unpredictable thrills of sky watching.

LOOKING EAST

8. Death

How Stars Die

A poet of Shakespeare's time was apt to say that his love for his beloved was not "sublunar." He meant that his affection was like the stars—constant and unchanging. According to the philosophy that prevailed in Shakespeare's time, change occurred only *below the Moon,* in our terrestrial world of jumbled elements and inconstant emotions. By contrast, the celestial realm above the Moon, like the poet's love, was fixed and eternal. Or so people believed. And then, to the consternation of poets and philosophers, in the autumn of 1572 a new star blazed in the constellation Cassiopeia. The star was well placed for viewing, high overhead in the evening sky. It was brighter than any other star, bright enough to be seen in broad daylight. All Europe was agog. But was this new star really above the Moon, in which case the philosophers were wrong and the heavens do change? Or was the "star" in Cassiopeia not a star at all but some sort of luminous object in the Earth's own atmosphere? There was one man in Europe eminently suited to answer the question—the highborn, brash, and gifted Dane, Tycho Brahe.

THE HUGE STAR ETA CARINAE BLOWS OFF LOBES OF GAS AS IT APPROACHES THE END OF ITS LIFE, EACH LOBE VASTLY LARGER THAN OUR SOLAR SYSTEM.

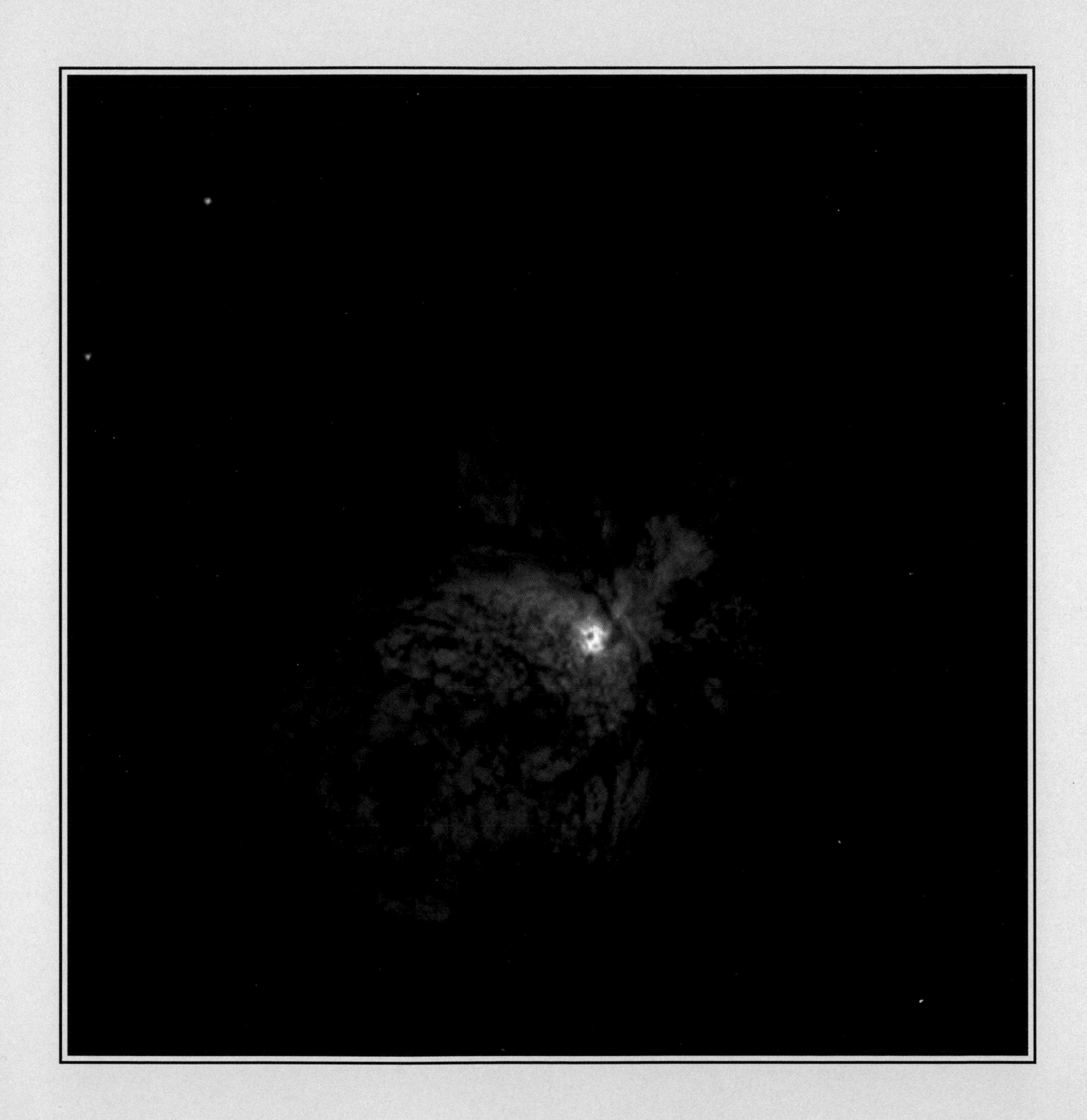

On the evening of November 11, 1572, Tycho was walking home to supper when he happened to glance up at the stars above his head. He saw the new star at once, the *nova stella*. Afraid to believe his own eyes, he asked his servants and neighbors to look into the sky and describe what they saw. They confirmed his observation. The star was real. Tycho was a talented astronomer, and he had recently built a fine new instrument (not a telescope; that device was still a few decades away) for determining the positions of stars. Carefully, he measured the angular separation of the new star from the stars of Cassiopeia. He continued these observations at different times of the night and throughout the winter as the star slowly faded. He also collected measured positions of the new star from other European observers. He was trying to determine the star's *parallax*, the apparent change in the position of an object when viewed from two different places (remember the finger in front of the nose and blinking eyes; see chapter 3). The object in Cassiopeia did not shift at all against the background stars when viewed from different places. From these observations Tycho knew that the new star was in the highest heavens, far beyond the Moon. The philosophers were wrong; the heavens do change.

We now know that the new star of 1572 was a supernova, the catastrophic explosion of a star too far away to have been visible before the detonation. The supernova of 1572 occurred at a crucial time in the history of astronomy. Thirty years earlier Copernicus had boldly suggested that the Sun, not the Earth, was the center of the universe. Forty years after the supernova Galileo, with the aid of the first astronomical telescope, forced a reluctant world to concede that Copernicus was right. The supernova of 1572 changed the supposedly unchangeable sky and helped tip the scales in favor of the brave new world that was emerging.

And as if one supernova was not enough, in 1604 another brilliant star appeared in the constellation Ophiuchus, not far in the sky from a rare conjunction of the planets Mars, Jupiter, and Saturn. That a "new star" should join an already remarkable gathering of planets was ominous. The apparition in Ophiuchus was on everyone's lips: Did it foretell the overthrow of the Turkish empire? A new European monarch? The Day of Judgment? On October 10, 1604, the new star was brought to the attention of Johannes Kepler, mathematician to the emperor Rudolph in Prague. If Tycho Brahe was the greatest observational astronomer of his time, Kepler was the greatest theoretical astronomer. He tracked the rise and decline in brightness of the new star over many months and collected data from other observers. He conclusively demonstrated the absence of parallax and therefore the celestial distance of the star. Once again, the "immutable" heavens of the philosophers had changed.

Both Tycho and Kepler published books on the new stars, adding to the growing debate about the true arrangement of the world. A few years later Galileo turned his telescope upon the heavens and discovered things that no bookish philosopher had dreamed of: spots on the Sun, mountains on the Moon, satellites of Jupiter, rings of Saturn (although not recognized as rings by Galileo), phases of Venus, and stars apparently without number twinkling beyond the limits of human vision. The heavens were positively *teeming* with variation and change, and out of all this turmoil modern astronomy was born. Before 1572, no bright supernova had appeared in Earth's sky for 400 years. In the thirty-two years that separated the two "new stars" of the Renaissance, the universe above the Moon (and below it, too!) was changed forever.

Naked-eye supernovas are rare, but astronomers frequently observe these celestial explosions with telescopes, in our galaxy and in other galaxies. Stars spangle the sky in uncountable numbers, and they are coming and going all the time. Low-mass stars like the Sun expire with modest violence. Five billion years from now, when the Sun has depleted the hydrogen fuel at its core, the core will collapse and the temperature will soar—from 10 or 15 million degrees Celsius to 100 million degrees. Then, an unstable struggle will begin between the squeeze of gravity and outward pressure of fusion, so long in balance. Deep inside the Sun, gravity will squeeze the core to ever hotter temperatures, and helium will begin to fuse into heavier elements—carbon and oxygen. The outermost layers of the Sun will swell, to envelop Mercury and Venus; eventually the star's tenuous surface will reach to the Earth or beyond. The surface of the Sun will then be cooler and redder, but its size will have grown so enormously that it will shine thousands of times more brightly than it does today. The Earth, of course, will be burned to a cinder, its oceans and atmosphere boiled away, and life (if it still exists) extinguished. As an unstable red giant, the Sun will puff off some of its outer layers, but eventually gravity will get a final grip on whatever material is left and squeeze it all back down into a hot, Earth-sized object called a white dwarf. Fusion will sputter out. The white dwarf will slowly radiate its residual heat until the star no longer shines. The Sun will be dead.

Big, high-mass stars live fast and die young. To support their greater weight, fusion at the core must proceed with fury, and high-mass stars use up their hydrogen fuel more quickly. Everything happens faster and more violently. At the end, the biggest stars die in shattering explosions called supernovas, scattering their substance into their galaxy.

In 1987, a supernova occurred 160,000 light-years from Earth, in a small companion galaxy of the Milky Way called the Large Magellanic Cloud. On the evening of February 23, Oscar Duhalde, night assistant at the Las Campanas Observatory in northern Chile, stepped outside for a look at the sky. He noticed an unfamiliar speck of light among the stars of the Large Magellanic Cloud. He did not realize it immediately, but he was the first person to discover a supernova with the naked eye in 383 years. Quickly, astronomers at Las Campanas recognized that the speck of light was a supernova, still growing in brightness, potentially the brightest supernova since the time of Kepler. The excitement of discovery flashed electronically around the world. As night rolled westward, astronomers and amateur star gazers in New Zealand and Australia took note of the blaze in the southern sky, and a frenzy of activity ignited. Observing programs were revised, observers went without sleep, plane tickets to the south were hastily purchased by astronomers in the north. Neutrinos pouring out of the exploding star were detected by instruments in a lead mine beneath the city of Kamioka in Japan, and in a salt mine under Lake Erie. The supernova grew in brightness, reaching a peak eighty-eight days after discovery, brighter than the stars of the Pleiades. Theoreticians scrambled to match rapidly accumulating observations with theories about how stars explode. Satellites in Earth orbit caught high-energy gamma rays from the supernova, confirming the idea that exploding stars forge new elements. The spectral signatures of molecules and radioactive elements were observed in the expanding supernova debris; the dying star was indeed seeding the universe with the elements of future stars and planets, just as the theoreticians had predicted.

In the following months and years, seminars and conferences were organized around the world, with heated debates about the meaning of data. Research grants were applied for, and astronomers competed for time on the southern hemisphere's great telescopes. Finally, as the explosion debris expanded and cleared, a blinking pulsar was revealed, the compact, rapidly rotating remnant core of the star. Also, the progenitor star of the 1987 supernova was identified on photographic plates made before the explosion, a hot, blue supergiant star in the unstable final throes of its life. It was everything the astronomers could have hoped for: a supernova close enough to be a subject of intense study, but far enough away to pose no danger to Earth.

What would be the effect on Earth if a star blew up in our neighborhood of the Galaxy? If a star as close as fifty light-years exploded, we would be in trouble. The blast would sweep the Earth with a burst of deadly gamma rays and X rays, and then a straggling wave of high-energy particles moving at nearly the speed of light. Several thousand years later, a slower-moving shell of exploded debris would arrive, also carrying deadly radiation. The solar system would lie within the danger zone for hundreds of years, and life on Earth would suffer a terrible blow.

Within fifty light-years of Earth there are approximately one thousand stars. The vast majority of them are smaller and less luminous than the Sun, and therefore unlikely to go out with a bang. Only a dozen stars in our fifty light-year neighborhood are larger and more luminous than the Sun. Vega is the star most like the progenitor star of the 1987 supernova. But Vega is smaller, less hot, and several thousand times less bright. If the astronomers' theories are correct, Vega will live one thousand times longer than the progenitor star of the 1987 supernova and die more sedately. Arcturus, Capella, and Pollux are approaching the end of their lives and have cooled and swelled to become orange giants, but they are not the kinds of massive supergiants that are expected to go supernova. In fact, there are no nearby supergiants, hot or cool. All things considered, our neighborhood seems a relatively safe place to be.

But neighborhoods change. Stars have independent motions in space, and over millions of years some of our closest stellar neighbors will move away and other stars will take their place. Meanwhile, the Sun itself is traveling through the Galaxy. All of this moving about complicates any attempt to estimate the chance of a nearby cataclysm. A reasonable guess is that a supernova might occur within fifty light-years of Earth once every few hundred million years. Mass extinctions in the fossil record may record some of these earlier traumas.

If you want to wait and watch for a supernova, keep your eye on the familiar star Rigel in the constellation Orion. It is similar to the progenitor of the 1987 supernova, a hot supergiant at that stage in its life when the balance of gravity and fusion can go dangerously askew. Fortunately, Rigel is at a reasonably safe distance from Earth—about 900 light-years. There are 5 million stars closer than Rigel, all smaller, which suggests how rare good supernova candidates actually are. Nevertheless, Rigel is 180 times closer than the star that blew up in February 1987. If Rigel should go supernova, it would become as bright as the Moon. On an otherwise dark night, you could read this book by its light. It would be the most spectacular celestial event in recorded human history.

Human life is short compared to the life spans of even the most short-lived stars, but our lives are nonetheless intimately connected to the deaths of stars. Nearby supernovas almost certainly caused traumas for life during the long course of terrestrial evolution. More intimately, our bodies are made of atoms forged in dying stars.

In the first minutes of the big bang, when primeval protons (the nuclei of hydrogen) fused to form heavier elements, only helium and a bit of lithium were created. At least, that's what the theoreticians tell us; it takes increasingly high temperatures to fuse the heavier elements, and before helium nuclei could start combining to make carbon, nitrogen, and oxygen, the universe had cooled to the point that this was impossible. The first stars were therefore made entirely of hydrogen and helium, and the only planets around those earliest stars were gassy hydrogen-helium giants like Jupiter and Saturn. There were no Earth-like planets made of silicon, oxygen, and iron; no carbon-based life.

Where, then, did the heavy elements come from? In the cores of massive stars, sufficiently high temperatures are available to forge all the elements up to iron. At first, during the normal lifetime of a star, hydrogen is fused into helium, and the star patiently builds a helium core. Eventually, hydrogen at the core is depleted, and fusion slows. Gravity gets the upper hand, the core collapses, the temperature soars, and helium nuclei are fused into heavier elements—carbon, neon, oxygen, silicon, and iron—by a complex sequence of nuclear reactions, all of them releasing energy. The core of a massive red supergiant star is structured like an onion: The outermost layer is hydrogen "burning" into helium; the next layer, helium fusing into carbon; moving inward, hotter and hotter layers of neon, oxygen, and silicon, each forging the elements of the next heaviest layer, with a kernel of iron at the very center. The protons and neutrons inside an iron nucleus are so tightly bound that no further energy can be extracted by fusion reactions. Iron is the end of the line. These final reactions happen with increasing rapidity: years, months, weeks, days. The core of a dying supergiant star is a factory for making heavy elements—on a runaway schedule.

But of what use are these elements for making future worlds if they stay locked up in the star? Fortunately, they have a way to escape. As silicon "burns" into iron, in the last and final fusion reaction, the superdense core collapses, bounces back, and the whole thing blows. Within seconds the star shatters. A supernova! The Milky Way Galaxy is littered with dead and dying stars wrapped in shrouds of expelled matter. We see them everywhere. The Hubble Space Telescope, in particular, has photographed a spectacular gallery of stars in their death throes, puffing off layers of their substance or blowing themselves to smithereens. Among the most dramatic Hubble images of dying stars is the supergiant Eta Carinae, 7,500 light-years away and one of the brightest and most massive stars in the universe: 4 million times more luminous than the Sun and 100 times more massive. Eta Carinae is ripe to become a pyrotechnic wonder. About 150 years ago, this super-supergiant star experienced a massive convulsion, blasting out two symmetric lobes of matter weighing as much as several Suns. The star (or perhaps it is two stars in a binary system) survived mostly intact, but the flare-up foretells more dramatic developments. The Hubble photograph of Eta Carinae evokes terrific violence (see photo for this chapter). The seething lobes of ejected matter are expanding at hundreds of miles per second. Each of them bears a striking resemblance to the fireball of a nuclear bomb explosion (which, in a sense, they are), although on a vastly different scale. Each of the star's fireballs is 10,000 times bigger than our solar system! Eta Carinae is about three times closer to Earth than the stars that blew up in 1572 and 1604, so it promises a sensational show if it goes supernova, brighter in our sky than Venus.

A supernova remnant that has been extensively studied by astronomers is called Cassiopeia A, about 11,000 light-years away. The shell of expanding debris is now 10 light-years across; planets of nearby stars, if they exist, have already been bathed in its fierce energy. As the stuff of the shattered star races outward, it collides with other matter in space. Shock waves bounce back into the remnant and heat it to tens of millions of degrees. The whole seething mass is an energetic source of X-ray radiation. The Chandra X-ray Observatory, launched by NASA into Earth orbit in 1999, detected in the remnant what no one had ever seen before: iron-rich knots of gas in the outer layers of the cloud, presumably blasted straight from the innermost core of the dying star. Silicon-rich blobs have also been observed, although not as far from the core as the knots of iron. The exploding core of the dying star apparently turned itself inside out as it hurled its substance into space.

So what are *you* made of? That's easy. Atoms of hydrogen, oxygen, carbon, and nitrogen mostly, with tiny bits of other elements such as calcium, phosphorus, and iron. And where did those atoms come from? "From my parents," a student will sometimes answer when I ask the question, and then realize immediately that can't be right; only an egg and sperm's worth of atoms came from our parents.

Then where? First, from mother's milk. Then from strained carrots and peas. Still later from cornflakes and pepperoni pizza. Our bodies take in atoms with our food and use them to build tissue and bone under instructions from the DNA that *was* our inheritance from our parents. And where did the carrots and peas and corn and cow get their atoms? From soil and air and water. And those? The atoms of the Earth were there within the presolar nebula of dust and gas out of which the solar system was born 4.5 billion years ago. And? And those atoms were blasted into the nebulae of the Galaxy from the heavy-element factories at the cores of massive stars that lived and died before the Sun and Earth were born.

Every atom in the universe, except the primeval hydrogen and helium from the fireball of the big bang, was forged in a star and hurled into space in a supernova explosion. Massive stars live only for tens or hundreds of millions of years; many generations of these stars were born and died within the Galaxy during the 5 or 10 billion years between the big bang and the birth of the Sun. With each supernova explosion, the amount of heavy elements in the universe slowly increased. Today, all together, they only amount to 1 or 2 percent of the total stuff of the universe, but that 1 or 2 percent includes the atoms of our own bodies.

Atoms (except for the few that are radioactive) endure forever. Let's end this story of dying stars by following a single oxygen atom along its path in space and time. It is born in the core of a red supergiant star in an outer arm of the Milky Way

Galaxy during the last few months of the star's life, deep, deep in the star's core, when a helium nucleus fuses with a carbon nucleus, binding nuclear particles more tightly and turning mass into energy—the star's last urgent hold on life. Then, catastrophically, the core of the star collapses, igniting frenzied nuclear reactions, blasting the oxygen atom out through the upper layers of the star. It races outward as part of an expanding fireball, eventually, after many millions or tens of millions of years, becoming a part of the interstellar medium. Perhaps it makes an alliance with a carbon atom to form a molecule of carbon monoxide, recognizable in the interstellar nebula by its spectral signature (if anyone is looking).

A billion years pass. Part of an interstellar nebula is pulled together by gravity to form a new yellow, midsize star, the Sun. The oxygen atom finds itself in the halo of comets that surrounds the new star, far beyond the outermost planets. Another billion years pass. A wandering red dwarf star pierces the comet halo, disturbing the cometary orbits, and our oxygen atom falls with its comet into the inner solar system of the Sun. After many looping orbits of the Sun, it collides with Earth. Now its vagabond adventures truly begin as the atom makes its way from atmosphere, to sea, to rock, to atmosphere again, forging now and again new molecular alliances, until finally, after a few billion years of wandering, it is taken up into the body of the night bird that sings as I stand beneath this starry sky—watching, waiting, knowing it may not happen during my lifetime, but always expecting that next supernova that will grace our sky, the next dying star, without whose death you and I and the night bird would not be here.

Wait until later in the evening, or into September, and you'll find two more zodiac constellations clinging to the southern horizon. But don't expect dazzling stars. Capricornus and Aquarius will be hard to see unless you have a clear, dark night. The two constellations are familiar only because the Sun comes this way during late winter on its climb back into northern skies. We read their names in newspaper horoscopes, but of course no astronomer any longer believes that the Sun's position among the stars on the day of your birth has any effect on your life or personality.

As we move into this empty quarter of the sky, we are looking down out of the spiral disk of the Milky Way Galaxy, which accounts for the paucity of bright stars.

The traditional figure of **Capricornus** is a Sea-goat—head and front quarters of a goat, tail of a fish. The origin of the figure goes back to Babylonian times, but of course the Greeks invented a story to explain it. The gods of Olympus were pursued to Egypt by the fearsome monster Typhoon, whose snakelike arms terminated in one hundred dragon heads. The god Bacchus, disguised as a goat, leaped terrified into the Nile, where he somehow acquired the tail of a fish. Eventually, the gods prevailed over Typhoon, and Zeus put fish-tailed Bacchus among the stars.

Aquarius, the Water Carrier, is another ancient constellation in this "watery" part of the sky. (Nearby are Cetus, the Whale; Pisces, the Fish; and Pisces Austrinus, the Southern Fish.) Three stars in Aquarius have names that mean "lucky." Thousands of years ago the Sun was among these stars when winter had passed and the gentle rains of spring brought new growth to fields and gardens. Good luck, indeed.

The "star" of stars in this part of the sky is **Fomalhaut** *(FOAM-al-ought)* in Pisces Austrinus. It stands alone, close to the southern horizon, almost as bright as Altair. It would be far better known to northern observers if it were higher in the sky.

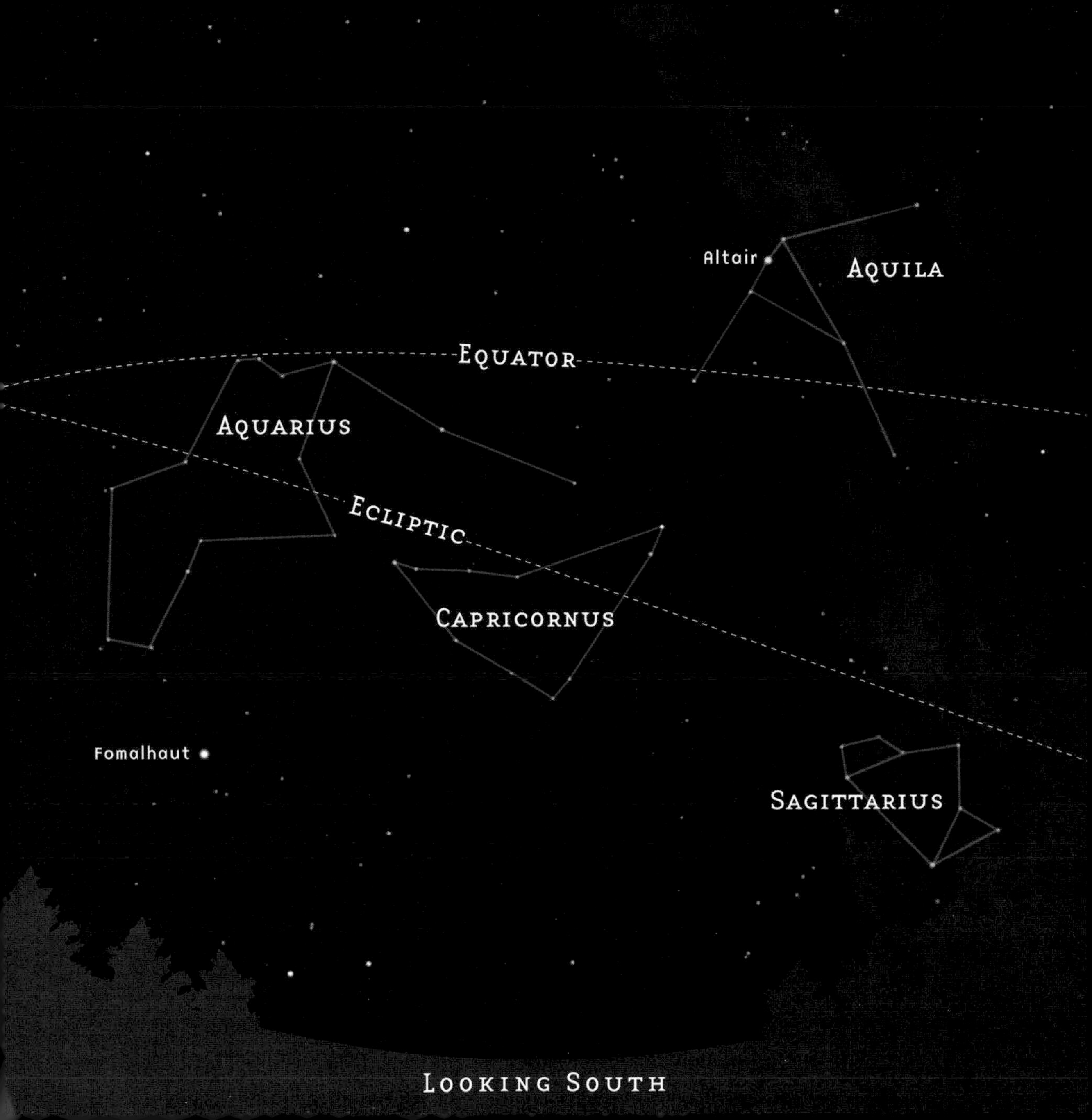

Altair

Aquila

Equator

Aquarius

Ecliptic

Capricornus

Fomalhaut

Sagittarius

Looking South

The name Capricornus is well known to students of geography from the **tropic of Capricorn**, the line on the Earth's surface parallel to the Equator that marks the Sun's deepest excursion into southern latitudes. The Sun stands vertically above the tropic of Capricorn on our winter solstice, about December 21. Similarly, the **tropic of Cancer** marks the Sun's northernmost excursion in the sky—on the summer solstice, about June 21. You will see from the map that when the Sun is in Capricorn today, it is *not* at its southernmost excursion, which actually occurs when the Sun is in Sagittarius. Because of the 26,000-year wobble of the Earth's axis—precession—the solstices and the equinoxes (where the Sun crosses the equator) have shifted by about one constellation since those ancient times when the tropics were given their names. More properly, the tropic of Capricorn should today be called the tropic of Sagittarius; likewise, the tropic of Cancer should today be called the tropic of Taurus. But we stick with the old names because no one has thought of any good reason to change them. In these days of satellite navigation systems and atomic clocks, no one pays much attention to the background constellations of the Sun.

One of Charles Messier's "fuzzy spots" can be found in Capricornus: M30, a globular cluster. You will recall that about one hundred of these ball-shaped swarms of stars are scattered above and below the disk of the Milky Way Galaxy. M30 is 40,000 light-years away, which places it somewhat farther away than the center of the spiral. With the map as a guide, try to imagine the actual three-dimensional location of M30 relative to the other stars. When we look toward Sagittarius, we are looking toward the center of the Milky Way spiral from our position about two-thirds of the way out toward the edge. We see M30 far beyond the nearby stars, below the galactic center, and a bit beyond. If you lived on a planet of a star in the cluster—and if your star was on the side of the cluster facing the Galaxy—imagine the view! Your entire night sky would be filled with a fabulous pinwheel of light. For such an observer, our own Sun would be an invisible speck of light lost among the one trillion stars of the Galaxy.

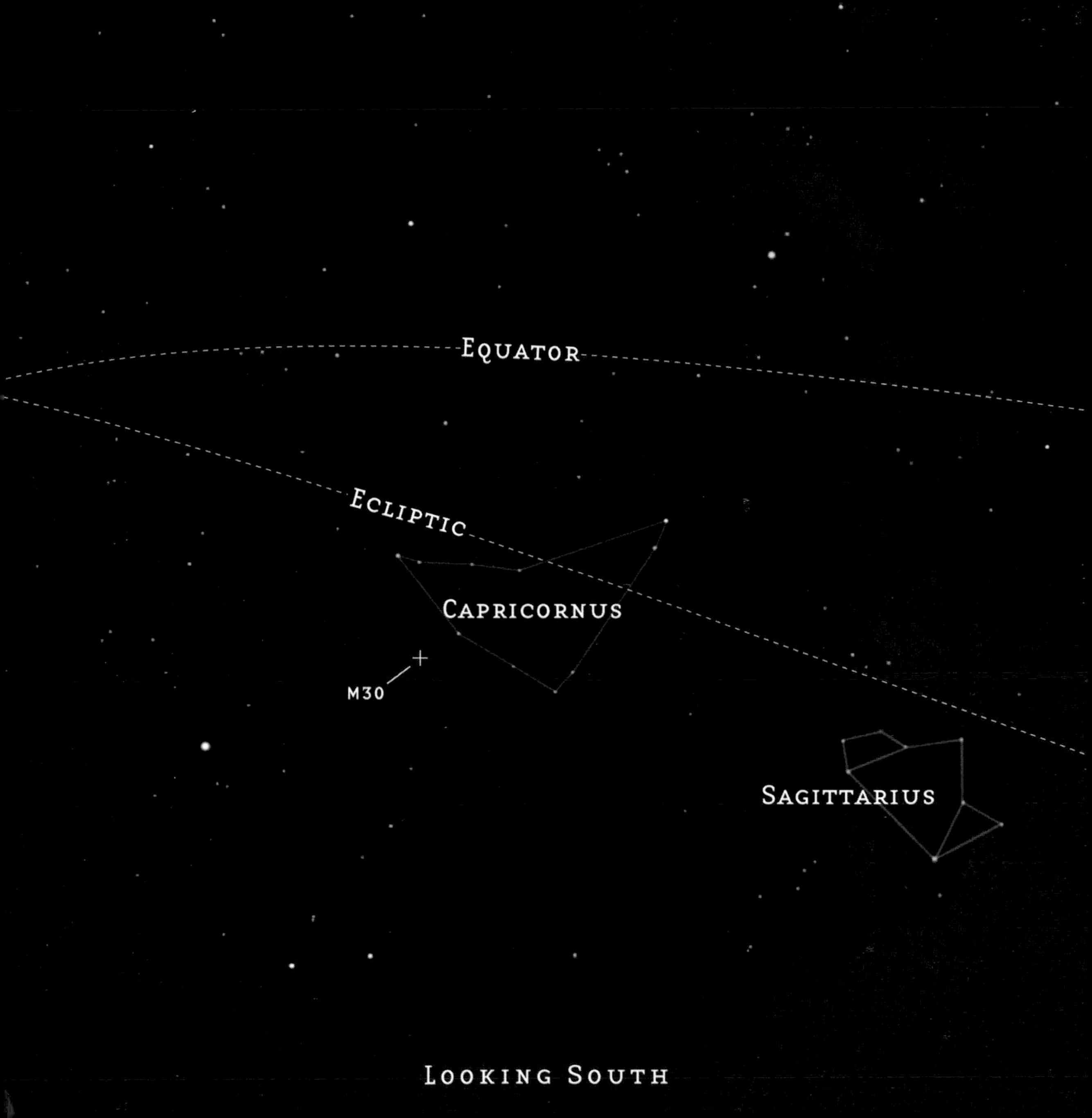

Equator

Ecliptic

Capricornus

M30

Sagittarius

Looking South

9. Vortex

The Milky Way Galaxy

Late summer is the best time for star watching. The Summer Triangle—Vega, Deneb, and Altair—is high overhead. In the north, Cassiopeia circles above the pole, turning its low *W* into a high *M*. In the south, red Antares burns, a cool giant star approaching the end of its life. And arching from north to south across the sky, the summer Milky Way—a river of luminous light, our spiral galaxy seen from the inside out.

When I was young, we thought ourselves citizens of the solar system. The sizzling, cloud-wrapped surface of Venus; the ice caps and "canals" of Mars; Saturn's jaunty rings—these were the horizons of our imaginations. Beyond Pluto, those one trillion stars in the whirling disk of our Milky Way Galaxy were too remote, too vast, to comprehend. The Milky Way was then the threshold of infinity. But—ah!—within the precincts of the solar system we found room enough for our fantasies. We visited the royal palaces of Jupiter's moons. We met and loved beautiful, green-skinned aliens from Planet X. We fought off refracto-

The Eagle Nebula, a giant star-birthing region in the Milky Way Galaxy, imaged by the Hubble Space Telescope.

ry Venusians who came in spaceships intent upon the colonization of Earth. And then, in the final decades of the twentieth century, those fantasies took on a kind of reality. We sent our own spacecraft—called Mariner, Viking, Voyager, and Magellan—on journeys of discovery to every planet (except Pluto) that circles the Sun.

The solar system has now become old hat, *terra cognito.* Voyages to the planets have become commonplace. The rings of Uranus are as familiar to us as our own backyards. Mercury's Sun-baked face is as ordinary as the Moon's. But that is as it should be. It is time to find new geographies to excite our imaginations, and astronomers are blazing the way *beyond* the yawning gulf that separates the solar system from our neighboring worlds. With fabulous new telescopes on Earth and in space, they are mapping the one trillion stars of the galaxy. Young sky watchers of the twenty-first century will be citizens of the Milky Way.

Every culture has had an explanation for the Milky Way, that band of milky light that circles the sky from Orion to Sagittarius and back again, arching high on summer and winter evenings, hanging low near the horizon in spring and fall. A bridge. A river. The pathway of the spirits of the dead. A circular cord of downy feathers. A track of cornmeal. In the National Gallery in Washington, D.C., is a painting by Tintoretto, titled *The Origin of the Milky Way,* that depicts the Greek view of things: Milk spurts into the sky from the breasts of the goddess Juno as she tries to nurse the lively infant Hercules.

Galileo's telescope put such fabulous stories to rest. In his little book *The*

Starry Messenger, he reported his observations: "For the galaxy is nothing else than a congeries of innumerable stars distributed in clusters. To whatever region of it you direct your spyglass, an immense number of stars immediately offer themselves to view." Not milk—nor feathers, nor cornmeal—but "congeries of innumerable stars." The evidence, as Galileo said, was indisputable. But astronomers for the next few centuries had other things on their minds besides those swarms of faint stars. Sun, Moon, planets, comets: their interests lay closer to home. Not until William Herschel turned his big new telescopes toward the stars in the late 1700s did the Milky Way again become an object of astronomical interest. Herschel plotted the positions of thousands of faint stars in three dimensions. (He assumed that all stars were of the same intrinsic brightness, and that therefore their apparent brightness was an indication of their distance.) We live in a disk-shaped cloud of stars, he concluded—a "grindstone," he called it—and he guessed that some of the other "fuzzy spots" in the sky were other Milky Ways of stars. We now know that he got it mostly right.

Herschel underestimated the size of our "grindstone" galaxy. He left out of his figuring the drifts of interstellar gas and dust that obscure distant stars. The astronomer who finally guessed the true size of the Galaxy was Harlow Shapley, who in the first decades of the twentieth century used the big sixty-inch telescope on Mount Wilson in California to study the globular clusters—ball-shaped congregations of thousands of stars that lie above and below the dusty disk of the Galaxy. Because the clusters are *outside* the disk, they can be seen at great distances without the effects of obscuring matter. Comparing variable stars (stars that change brightness) in the clusters to similar nearby variable stars, Shapley worked out the distances to the clusters. He assumed that the center of the distribution of globular clusters was also the center of the Milky Way Galaxy, and that therefore the Galaxy was tens of thousands of light-years wide—a spinning pinwheel of stars with the globular clusters buzzing about its center like bees. The Galaxy was so big that Shapley could not believe that the spiral nebulae—those other spiral-shaped "fuzzy spots" that sprinkle the sky—were equal objects. The universe would have to be absurdly large to contain so many Milky Ways, he thought. The Milky Way Galaxy *was* the universe, he concluded, and the spiral nebulae must be different kinds of objects, much smaller, and *inside* the Galaxy among the stars.

Earlier in this book, I mentioned biologist Richard Dawkins's Argument from Personal Incredulity: *If it seems impossible to me, it must be wrong*. It is a flawed argument, of course. The universe never ceases to confound our expectations, as in this case it confounded Shapley's. Not long after he had come to his conclusion about the nature of the spiral nebulae, a letter arrived at his office in the East from the Mount Wilson Observatory. On reading the letter, Shapley said to a colleague, "Here is the letter that has destroyed my universe." The missive was from Edwin Hubble, who had just determined the distance to the one of the spiral nebulae—the one in Andromeda—using the method of variable stars. A new, bigger, one-hundred-inch telescope constructed on Mount Wilson—the world's largest—had made it possible to resolve the Andromeda spiral into individual stars, which could then be compared to stars closer to home. The spiral was nearly *1 million* light-years away, said Hubble. (We now know the distance is closer to 2 million light-years). So the spiral nebulae were indeed other Milky Ways. And they sprinkle the space of the universe as far as Hubble's telescope could see. The universe was bigger—vastly bigger—than even Shapley had dared to think.

The physician-essayist Lewis Thomas once bluntly asserted: "The greatest of all the accomplishments of twentieth-century science has been the discovery of human ignorance." It is an odd, unsettling thought that the culmination of our greatest century of scientific discovery should be the confirmation of our ignorance. How did such a thing come about?

When we thought we lived on a globe at the center of a sphere of stars *just up there* on the dome of night, we could easily believe that a complete inventory of the universe's contents was possible. The cosmos was proportioned to a human scale, created specifically for our domicile. Presumably, since the universe was made for us, it contained nothing beyond the ken of our senses. But the telescopic observations of Galileo, Herschel, Shapley, Hubble, and many others revealed stars in uncountable numbers, stars that served no apparent purpose in the human scheme of things since they could not be seen by human eyes. The Hubble Space Telescope's Deep Field Photographs (see photo for chapter 1) show galaxies as numerous as snowflakes in a storm, each with uncountable planets, strange geographies, perhaps biologies and intelligences. To live in such a universe is to admit that the human mind will never know everything. The philosopher Karl Popper wrote: "The more we learn about the world, and the deeper our learning, the more conscious, specific, and articulate will be our knowledge of what we do not know, our knowledge of our ignorance. For this, indeed, is the main source of our ignorance—the fact that our knowledge can be only finite, while our ignorance must necessarily be infinite."

Are we psychologically prepared to be citizens of the Milky Way? Some of us are frightened by the vast spaces of our ignorance and seek refuge in the human-centered universe of the ancients. Others are exhilarated by the opportunities for further discovery, for the new vistas that will surely open before us. The discovery of our ignorance should not be construed as a negative thing. Ignorance is a vessel waiting to be filled, permission for growth, a ground for the electrifying encounter with mystery.

Ironically, we now know more about nearby galaxies than we do about our own. The problem, of course, is that we are *inside* the Milky Way, a spiral of one trillion stars, on a tiny planet orbiting a single star, and we have only a hazy notion of the shape of that stellar mass. In Herschel's view, we were smack dab in the middle of a "grindstone" of stars—a circumstance which was so suspiciously anthropocentric (human-centered) that he came to believe that his view must be incorrect. A century later, Shapley got it right; we are somewhere off to the side of a pancake-shaped cloud of stars, but still more or less centrally embedded in the pancake. The space between the stars is full of dust and gas, which makes it difficult for us to see very far into the pancake. Up and down *out of the pancake* the view is fairly clear, which is why we can see other galaxies in those directions. Once Hubble convinced us that the spiral nebulae are other Milky Ways, we could assume that if we went outside of the Milky Way and looked back, it too would look like a spiral. But of course, getting out of the Galaxy is impossible. For a spacecraft (of the kind we could build today) to reach the upper or lower edge of the Milky Way and escape from the disk would require millions of years of travel. Leaving the Galaxy is just not in the cards, at least not with any technology we can currently imagine. OK, then let's make radio contact with members of an intelligent civilization in another galaxy—the "nearby" Andromeda Galaxy, say—and ask them to transmit an image of our Milky Way Galaxy. But, of course, the Andromeda Galaxy is 2 million light-years away, so a round-trip message and response would take 4 million light-years. We are on our own, and we have to map the Milky Way Galaxy *from inside* as best we can.

It turns out that "as best we can" is getting better all the time. Telescopes on the ground and in Earth orbit probe the Milky Way in every part of the electromagnetic spectrum: radio, microwave, infrared, visual, ultraviolet, X ray, gamma ray.

Computers, too, are telescopes of sorts. They allow astronomers to calculate the dynamics of gas and stars in faraway regions of the Galaxy; the results of the calculations can then be compared with observations, allowing astronomers to turn static observations into history. As the new millennium begins, our maps of the Milky Way Galaxy are almost as good as our maps of the solar system.

Imagine, then, a journey to the center of the Milky Way, a place hidden from our visual inspection by the dust and gas that fill the space between the stars. Radio waves and X rays penetrate the mists, revealing hints of structure. Not so long ago we believed that the Milky Way was a normal spiral galaxy, with starry arms pinwheeling out from the very center. But no! It is a "barred" spiral. Two spokelike arms extend straight out from the center of the Galaxy on opposite sides, then bend to wrap their way to our neighborhood some 26,000 light-years from the center. As the Galaxy rotates, the bars pump gas toward the center of the Galaxy (this can be calculated with computers and observed with telescopes). Monstrous clouds of gas fall inward along wildly dancing orbits, colliding energetically, shining with a fierce high-energy radiation. Now we approach the central bulge, or nucleus, of the Milky Way, a region as small in the Galaxy as a grapefruit seed at the center of a dinner plate. Here the pressure of the gas builds, as if in a pressure cooker, until it explodes in bursts of out-rushing wind. One might think that with so much gas at such high pressure, stars would spring prodigiously into existence, for it is out of the interstellar gas that stars are born.

But the rate of star formation in the nucleus is modest. The violence, the stirring, the intense magnetic fields inhibit gravity from gathering the stuff of new stars.

As we draw closer to the Galaxy's core—to the very center of the grapefruit seed—things become strange indeed, not quiescent, as one might expect at the eye of a storm, but more violent. Strong magnetic fields squirt jets of gas up and down out of the Galaxy, and supernova explosions trigger massive disruptions. At the core of the Galaxy, massive stars flash into existence, blaze with a quick intensity, then explode. Stars collide or tear at each other with gravitational claws. The density of stars is one million times greater than in the arms of the Galaxy. If there are inhabitants of planets here at the center—which seems unlikely given the terrible violence of the neighborhood—their night sky shines with daylike brilliance. At the very center of the Milky Way lies a black hole with the mass of one million Suns—the dark and violent heart of the Galaxy. (A black hole is a spherical region of space where gravity is so strong that not even light can escape. Nothing can break away from the fierce grip of a black hole.) This black hole may have had its origin early in the universe's history when a huge gas cloud near the center of the young Galaxy collapsed under its own gravity. At first, the black hole may have had a mass of no more than a few hundred thousand Suns, but it subsequently gobbled up gas or stars that drifted too close. Even today, episodically, clumps of gas stream toward the black hole, radiating fits of energy outward before disappearing forever into oblivion.

Astronomers have glimpsed stars in tumultuous orbits around the Galaxy's central black hole. Calculations suggest that every 10,000 years or so one of these stars will wander too close to the black hole and be ripped to shreds. As the remnants of the star plunge into the monster's maw, its doomed matter shines more brightly than a supernova, briefly outshining the entire galaxy. *Every dark pool should have its monster. Every dark pool should be haloed with light.* The Milky Way is quiescence and violence, radiance and darkness, life and death—a whirlpool of continuous creation and destruction.

Out in our neighborhood of the Galaxy, just on the inside of a spiral arm, two-thirds of the way from the center, things are quieter. Stars are born; stars die. The Sun travels on its long uneventful journey about the galactic center, taking 200 million years for one revolution; the last time the Sun was where it is now, dinosaurs ruled the Earth. Above and below the plane of the Galaxy are the globular clusters, glittering balls of tens of thousands of old stars, in great looping orbits about the galactic center, occasionally diving down through the disk of the Galaxy and out the other side. Within a cluster, stars buzz in their own orbits about the cluster's center. (We do not directly observe this frenzy of activity, any more than we can watch the wheeling of the spiral; it takes place on a timescale much longer than the human life span.)

Just as the Galaxy has an invisible darkness at its core, it is also surrounded by an invisible dark halo, an all-enclosing cocoon of mysterious matter that may contain as much as 95 percent of the Galaxy's mass. If the cocoon can't be seen, how do astronomers know it's there? By its gravitational effects on visible stars and on nearby galaxies. What is it made of? Ah, that's the riddle of the *dark matter*. Ancient dwarf stars? Exotic subatomic particles? Black holes? No one knows. Other galaxies have similar dark matter halos. It is a measure of our ignorance that most of the *stuff* of the universe is invisible and unknown. Eventually the riddle of the dark matter will be answered.

In the meantime, astronomers have sketched out roughly the distribution of the one trillion stars and the banks and drifts of dust and gas in the Galaxy. The Milky Way is no longer the threshold of infinity; we have plumbed its depths, searched its hidden crannies, and glimpsed worlds more wonderful and bizarre than anything we know in our solar system. In the greater universe of the galaxies, the Milky Way is our home.

Where did it come from, this vortex of stars with the monstrous core and hidden halo? And what will be its fate? Astronomers have seen hints of the Milky Way's history in populations of stars within the Galaxy that have different apparent collective motions. The Milky Way seems to have been forged from the merging and amalgamation of smaller galaxies. This process of amalgamation may not be over yet. Nearby minigalaxies, rich in hydrogen gas and dark matter, continue to fall into the Milky Way, generating waves of star birth. These minigalaxies may be typical of the first galaxies in the universe, with giants like the Milky Way and the Great Andromeda Galaxy growing by gobbling up the dwarfs.

But something even bigger is in store for the Milky Way, bigger than these minicollisions. Although the galaxies in general are racing apart from one another with the expanding universe, local groups of galaxies do not necessarily disperse. The Andromeda Galaxy—about the size of our own—is coming our way. Or rather the Andromeda Galaxy and the Milky Way Galaxy are closing the gap between them at 300,000 miles per hour. At that pace they will collide within 3 billion years. At first, the two galaxies may whirl past or through each other, dragging out long gravitational tails. The Milky Way will be distorted out of recognition. But then the gravitational attraction of the galaxies's dark-matter halos will cause them to coalesce. When everything settles down after 1 billion years or so, the combined ex-spirals may have created a giant egg-shaped galaxy blazing with newborn suns. These new star systems will be rich with heavy elements and shepherd Earthlike planets. Allow a few more billion years for evolution to spin out complex forms of life, and the glory days of intelligent life in our part of the universe might truly begin in earnest.

IV. Autumn

As we move into autumn, a vibrant celestial drama unfolds in the night sky. Stretch out on your back, facing north, and watch characters of Greek mythology parade overhead. Here is the cast of characters:

Cassiopeia *(Kass-see-oh-PEE-ah),* queen of Ethiopia, sitting on her throne. Look for a lopsided *M* of stars, directly opposite Polaris from the Big Dipper.

Cepheus *(SEE-fee-us),* her husband, king of Ethiopia. Look for an upside-down "house" of stars, below the *M* of Cassiopeia, not as prominent as the stars of the queen.

Andromeda *(an-DROM-eh-dah),* their beautiful daughter, a sweeping vee of stars curving down toward Cassiopeia from a corner of the Great Square (see below).

Perseus *(PER-se-us),* hero, slayer of the gorgon Medusa with the snaky hair, so ugly that anyone who observed her directly turned to stone. Perseus chopped off her head while looking at her reflection in his shiny shield. Look for a ragged vee of stars to the east of Cassiopeia. One of these, **Algol** *(AL-gall),* the Ghoul, is the baleful head of Medusa.

And **Pegasus,** Perseus's winged horse, best recognized as a **Great Square** of stars high overhead.

The story: Cassiopeia boasted that her daughter, Andromeda, was more beautiful than the sea nymphs. The jealous nymphs protested to the sea-god Poseidon, who sent a sea monster Cetus (see the last map of the book) to ravage the coast of Ethiopia. Cepheus asked an oracle how the devastation could be avoided. His kingdom could be saved only if he sacrificed his daughter to the monster, he was advised. Cepheus was torn between love of Andromeda and love of his people, but at last he ordered Andromeda chained to a rock on the coastline for the monster to devour. As Cetus approached the terrified girl, who should come flying overhead on his winged steed but Perseus, returning from slaying Medusa. He saw the terrible event that was about to unfold, took the gorgon's head from the sack in which he carried it, and, without looking himself, held it out for the gaze of Cetus—who promptly turned to stone and sank to the bottom of the sea. Perseus released Andromeda from her chains and, smitten by her beauty, asked Cepheus for her hand in marriage. The two lived happily ever after.

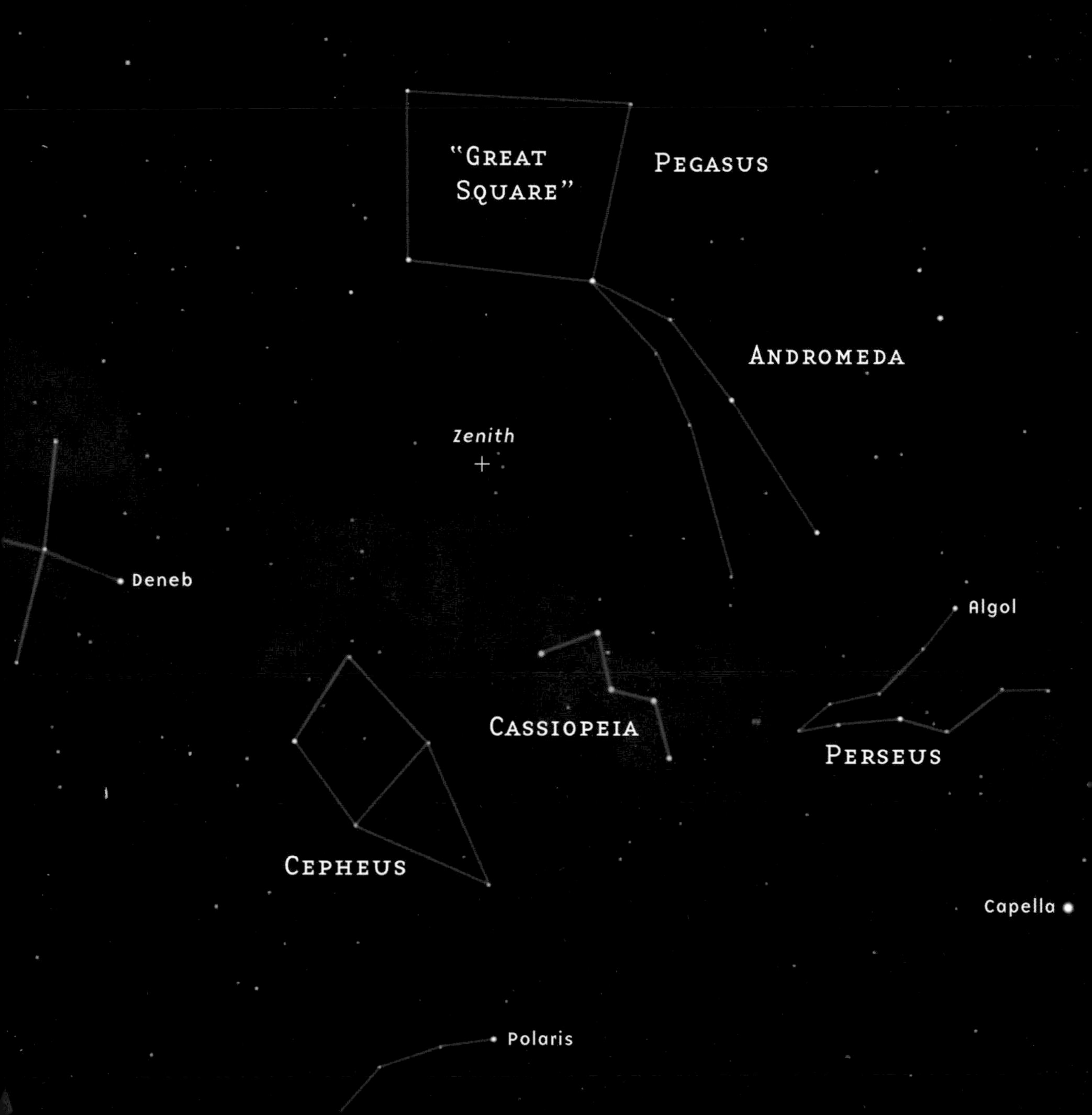

"Great Square"

Pegasus

Andromeda

Zenith

Deneb

Algol

Cassiopeia

Perseus

Cepheus

Capella

Polaris

Cassiopeia lies astride the Milky Way, and the constellation harbors many telescopic riches: clusters of stars recently born from the Galaxy's store of dust and gas. A typical cluster might contain one thousand stars in a region of space only tens of light-years across, including many hot blue giants. The night sky from a planet in such a cluster must be spectacular, but the stars are young and presumably there has not been time enough for conscious life to evolve.

Young clusters are violent places, where blue giant stars blow themselves to smithereens. The biggest, most massive stars burn their fuel at such a furious pace that their lifetimes are measured in millions of years, not billions. When they go, they go with a bang. A supernova appeared in Cassiopeia on November 11, 1572, and for a brief time was brighter than Venus, visible even in daylight. Slowly it faded, changing color from white to yellow to reddish, until it disappeared from view about two years after its sudden appearance.

As we have seen, this "new star" was observed by the great Renaissance astronomer Tycho Brahe, for whom it is named: **Tycho's Star.** It was a miracle, he thought, the first such event in the history of the world, and like everyone else he wondered what it meant. He was wrong, of course—about the miracle. Supernovas occur in our galaxy every few hundred years, and another would blaze out not many years later in 1604. This was the most recent, and if history is a guide we are overdue for another celestial pyrotechnic.

Algol, the Ghoul, in Perseus, had a baleful reputation long before its association with Medusa, probably because of the star's mysterious variation. Every three days it has a sudden dip in brightness, briefly surrendering equality with the other bright star in Perseus, Mirfak. Does Medusa blink her eye? Algol is an **eclipsing binary,** two stars—one hot and bright, one cool and dim—in close orbit around each other. Every three days the dim star moves in front of the bright one, eclipsing its light. When the bright star moves in front of the dim one, the dip in brightness is not so dramatic.

Algol
Tycho's Star
1572
Mirfak

10. Myriad

The Universe of Galaxies

These days, astronomers build their giant telescopes in the Chilean Andes or on Hawaii's extinct volcano Mauna Kea, high, clear places thick with stars. No one would think of siting an observatory in Ireland—with its wet-rag skies, its one clear night in ten. But things were different a century and a half ago. Travel to far-off places was not so easy, and if you happened to live on a grand Irish estate, and you had a passionate interest in the sky, especially in the faint fuzzy nebulae that sprinkle spring and autumn skies—well, then you made do with whatever occasional starry night that nature provided. During the 1840s, William Parsons, third Earl of Rosse and master of Birr Castle in the center of Ireland, constructed what was then, and would remain for three-quarters of a century, the world's largest telescope.

Lord Rosse's telescope was a monster, an iron-hooped cylinder more than six feet in diameter hoisted up between massive Gothic walls, with ladders and viewing galleries—a Big Bertha howitzer aimed at the stars. Visitors to the castle liked to have their pictures taken (by the earl's wife, Mary, a pioneer amateur photographer) standing in the gaping throat of the tube. The heart of the instrument

The majestic spiral galaxy NGC 4414, located about 60 million light-years from our Milky Way Galaxy.

was the "Great Speculum," a metal mirror, six feet in diameter and weighing four tons, that gathered the light of distant nebulae and brought it to a focus. The disk was cast in a special furnace constructed in the castle moat, fired with turf from local bogs. Polished to a high brilliance, it was a wonder of Victorian science.

Astronomers from as far afield as the United States, Australia, and Russia came to Birr to see Lord Rosse's leviathan eye on cosmic deeps. One wonders how many of them managed to get a look at the stars or went away instead cursing the Irish weather. Of course, Rosse himself had the leisure to wait for clear nights, and when they came he went exploring. Of all the things he saw with his wonderful instrument, he is best known for his discovery of the spiral nebulae.

In 1845, he observed a face-on pinwheel of light in the constellation Canes Venatici, under the handle of the Big Dipper. A luminous blur at this location was first cataloged by the French astronomer Charles Messier in 1773. Later, the astronomer Sir John Herschel, son of William (of Uranus fame), examined the blur with an 18-inch telescope and detected a "very bright round nucleus surrounded at a distance by a luminous ring." But it was Rosse's 72-inch colossus that revealed the unlikely and mysterious spiral. The discovery of other celestial pinwheels soon followed. At first the "spiral nebulae" were thought to be new solar systems in the process of formation, not so far away, just out there among the nearby stars. As we have seen, it wasn't until the construction of the 100-inch telescope at Mount Wilson in California, in 1923, that the spirals were determined to be other "island universes," or galaxies.

Today, we call the object in Canes Venatici "the Whirlpool Galaxy." Modern photographs show a dazzling double coil of tightly spiraling arms sprinkled with stars and streaked with dark lanes of dust. The Whirlpool Galaxy is 15 million light-years distant and contains at least 100 billion stars. It is very much like our own Milky Way Galaxy, from which it is receding at 340 miles per second as part of the general expansion of the universe. Tonight it will be 30 million miles farther away than last night, more than one trillion miles farther than when Rosse first sketched its pinwheel form.

All galaxies, except those in our Local Group, are speeding away from us, hurtling into darkness, stretching and dimming their light, pulling the universe thin like taffy. Blasted outward by the violence of creation, they may continue to recede forever, or, in the apparently unlikely event that their mutual gravity is sufficient to halt their recession, they will snap back like the elastic-tethered balls of paddle bats to re-create the big bang.

All those galaxies racing away into the night are not inert objects, like stones skipped on a pond. Surely they snap, crackle, and pop with life. Surely the stars of those other galaxies shimmer in their own green auras, their planets puffing with spores. The Whirlpool Galaxy is a prayer wheel spinning out our supplications to the dark abyss, born in fiery brilliance, heading toward a still uncertain fate. To celebrate the 150th anniversary of Lord Rosse's discovery of the spiral nebulae, a massive fund-raising drive was undertaken in Ireland to restore the great telescope at Birr and show it off, complete with the original staircases, chains, pulley systems, and viewing galleries, as part of a historic science center. The original "Great Speculum" remains in the Science Museum at Kensington, in London. When last I saw it there, it reposed in its reflective glory on a horizontal stand, like a magical Round Table from Camelot. It is replaced at Birr by a replica. Today, visitors to Ireland can visit Lord Rosse's estate and see the instrument that brought the spiral galaxies within our ken.

Only three galaxies besides our own are visible to the naked eye. The Large and Small Magellanic Clouds are prominent objects in the southern hemisphere, patches of hazy light like broken-off pieces of the Milky Way that drifted into other parts of the sky. These nearby dwarf galaxies, of irregular shape, are bound to our own galaxy by gravity. The Large Magellanic Cloud is 160,000 light-years away (about twice the diameter of the Milky Way spiral). The Little Magellanic Cloud is nearly 200,000 light-years away. Think of an orange and a lemon hanging a few feet below a dinner plate (or better, two dinner plates placed together face-to-face), and you have a general notion of the size and location of the Magellan galaxies relative to our own. The first European navigators to sail around Africa's Cape of Good Hope called these blurs the Cape Clouds, but later they came to be named for the great circumnavigator who sailed through southern seas. Of course, every indigenous people of the southern hemisphere had names for these prominent objects; Polynesian islanders called them the Upper and Lower Mist.

In spite of their visual prominence, the Magellan galaxies are not in fact the nearest ones to our own. In the mid-1990s, a closer dwarf galaxy was discovered in the constellation Sagittarius, mostly hidden by the obscuring matter of the Milky Way. It is smaller than the Magellan galaxies (think of a grape just at the edge of the dinner plates). If the Sagittarius dwarf were on *our* side of the Milky Way Galaxy, instead of all the way across the center, it would be a prominent object in our sky. Its fate is surely to be gobbled up by the Milky Way—one more morsel in the ongoing feast that has fattened our galaxy's bulk. Another half dozen or so dwarf galaxies, visible only with telescopes, lie within 1 million light-years of the Milky Way, more future snacks for our galaxy.

In the northern hemisphere, we have only one naked-eye galaxy, the Andromeda Galaxy (see next star map), a magnificent spiral about the size of the Milky Way, 2 million light-years away. As we have seen, Andromeda is moving our way, and sooner or later it may merge with the Milky Way. You will need a clear and very dark night to see it—a hazy patch of light along Andromeda's flowing gown. What one sees with the unaided eye is only the bright central region of the galaxy, an ellipsoidal mass of old red and yellow stars; the spiral arms of the galaxy, with their diffuse banks of gas and dust and hot young blue stars, are too faint to be visible. The Andromeda Galaxy is actually four times bigger than the Moon in our sky; if we could see its full extent, it would make a glorious sight wheeling among the stars. The only other spiral galaxy in our Local Group is the Pinwheel Galaxy in the constellation Triangulum, not so far away in the sky from the Andromeda Galaxy on the other side of Andromeda's body. Observatory photographs reveal a magnificent face-on spiral.

Beyond the Local Group, galaxies fill space like the motes of dust that one sees in light streaming through a window into a darkened room. I once had the privilege of visiting the observatories on Siding Spring Mountain in New South Wales, Australia. The setting for this great scientific facility is spectacular. On every side of the observatories rise the jagged volcanic peaks of the Warrumbungle Mountains, mantled with gum trees. In the grassy valleys at the base of the mountains kangaroos and emus graze at dawn and sunset. Beyond the mountains the featureless outback reaches to the far horizon.

One of the smaller domes on Siding Spring Mountain houses the British 1.2-meter telescope, which was then engaged in a systematic photographic survey of the southern sky. The photographs are made on emulsion-coated glass plates. The plates are thin— not much thicker than a thumbnail—so they can be bent to match the curved focal surface of the telescope. Each plate is about the size of a newspaper page. A typical exposure lasts about one hour, during which time the telescope moves with extreme accuracy to compensate for the turning of the Earth. Each of the glass-plate photographs contains between 1 million and 10 million visible images. The images are sharp spots and fuzzy spots. The sharp spots are stars in our own Milky Way Galaxy. The fuzzy spots are mostly other galaxies, other island universes that contain as many as one trillion stars each. About half of the images on any single plate are galaxies.

I examined several contact negatives with a magnifier. (The plates themselves are far too precious to be made available to visitors.) In the magnifier, the brightest of the fuzzy spots became spiral galaxies of dazzling detail. Many of the fuzzy spots were interacting galaxies, two or more great star systems locked in a spiral-distorting gravitational dance. It would require almost 1,800 of these photographs to cover the entire sky, and on each plate there are recorded as many as a million galaxies! With the magnifier, I examined in a few minutes more worlds than my mind was capable of comprehending.

What does it mean, this extravagant profusion of worlds? I had gone halfway around our tiny planet to visit Siding Spring Mountain; it seemed an enormous distance. Yet if I were traveling in a Quantas jumbo jet, it would take me *more than one trillion years* to reach the *nearest* of the galaxies on the photographs. Every galaxy fixed on a photographic plate has become a permanent part of the human imagination. Through the agency of the telescopes on Siding Spring Mountain, and others like them throughout the world, the human mind has gone out to embrace the distant galaxies. We may be physically small compared to the galaxies, but our minds can be as large as the universe. As I examined the contact prints, this thought occurred to me: We are almost certainly not the cleverest thing the universe has yet thrown forth, but we are certainly not small. Our imaginations are billions of light-years wide. We can justifiably say with Shakespeare's Miranda: "O, wonder! How many goodly creatures are there here! How beauteous mankind is! O brave new world that has such people in't!"

The photographic prints I examined at Siding Spring Mountain each record a part of the sky I might cover with my palm held at arm's length, and on them are millions of tiny dots and blurs. In the magnifier, some of the blurs appeared as perfect little spirals, whirling and colliding in astonishing numbers. Each of the Hubble Space Telescope Deep Field Photographs (see chapter 1 photograph, and chapter 3) shows a few thousand galaxies in an area of the sky that could be covered by the intersection of crossed sewing pins held at arm's length—one hundred times more galaxies per square degree of sky than are visible on the Siding Spring plates. Think of the Milky Way Galaxy as a dinner plate. The next closest spiral galaxy—the Andromeda Galaxy—is another dinner plate across the room. On this scale, the nearest galaxies visible in the Hubble Deep Field Photographs are dinner plates about one mile away. The faintest galaxies in the Deep Field Photographs are dinner plates more than twenty miles away, at the very edge of space and time.

Can we see even deeper? The Next Generation Telescope, the Hubble's successor, is on the drawing boards. Meanwhile, ground-based instruments are getting bigger and better. Powerful computers adjust the shape of the mirrors of these instruments instant-by-instant to compensate for distortions of the atmosphere. By the time the Next Generation Telescope opens its eye on the cosmos, astronomers will already understand much more about the earliest days of galaxy formation.

Two mysteries that remain to be unraveled are the natures of dark matter and quasars. Dark matter, it seems, was there from the beginning, accounting for at least 90 percent of the universe's mass, and directing by its gravity much of what happened as the galaxies formed. Quasars are intensely energetic objects, much smaller than galaxies, that were far more common in the early universe than they are today. Most astronomers believe the quasars are the formation of massive black holes at the centers of the large galaxies as those galaxies were being born. Many or all of the large galaxies may have gone through a quasar stage lasting tens of millions of years. The black hole at the center of our own Milky Way Galaxy almost certainly had its origin in a quasar episode (or episodes) billions of years ago.

As I write this chapter, in the spring of 2000, the most distant object yet observed in the universe, with the Hubble Space Telescope, is a galaxy 13 billion light-years from Earth. Because the galaxy—dubbed "Sharon" after the sister of one of its discoverers—is so far away, and because light takes time to travel the distance between us, we see it as it existed when the universe was only 5 percent of its present age, not long after the first galaxies and stars condensed from the primeval matter of the big bang. Detection of such distant galaxies is a considerable technical achievement, not only because they are so faint and far away. Because these galaxies share in the universe's expansion, their light is stretched into parts of the spectrum that are absorbed by interstellar dust and gas that lie along our line of sight and by the Earth's atmosphere. Only with a telescope above the atmosphere is detection possible at all.

If we could visit the Sharon Galaxy *in the era in which we observe it*, it would be a lively place, roiling with primeval energy, exploding with the violent deaths of hot massive stars. But nowhere among the worlds of that galaxy would we find sperm, egg, wing, flower, feather. Not only were the elements of life unavailable, but not enough time had elapsed for complex life forms to evolve. The Sharon Galaxy presumably still exists somewhere in the universe today, 13 billion years older than it appears in the Hubble photograph. And what a different sort of galaxy it must be now! Countless generations of massive stars have lived and died in the galaxy during the ensuing eons. As stars burn, they fuse heavy elements from hydrogen and helium. When stars die explosively, they make still more heavy elements and spew those atoms into the interstellar nebulae out of which new generations of stars will be born. As time passed in the Sharon Galaxy, more and more heavy elements enriched the primeval gas mixture, until by now the most recent stars presumably contain a few percent of atoms like carbon, oxygen, and iron, and shepherd small planets made of rock, metal, water, and air.

Perhaps on one of those planets intelligent life has evolved and is looking our way. What would these beings see? They would see the ancestor of our present Milky Way Galaxy as a dot of light on the threshold of time, 13 billion years in the past—a galaxy without Earth-like planets, without life, without intelligence. They would see our galaxy in its infancy, hot and lively, growing by galactic cannibalism, and stewing up the atoms that will someday compose the Sun, the

Earth, the Hubble Space Telescope, and the bodies of the men and women who use that telescope to explore cosmic history. In other words, our galaxy appears to Sharonites—if such exist—more or less as their galaxy appears to us.

Can we observe the Sharon Galaxy *as it exists today?* Impossible. We can see that other galaxy in our space, but not in our time. We are sequestered in our present by the finite velocity of light. However, we have reason to believe that the laws of nature are the same throughout the cosmos, and that our galaxy is not untypical. So, in a sense, that dot of light 13 billion light-years away is "us"—or a galaxy very much like what our galaxy might have been like then. The Hubble Space Telescope is humankind's most effective time machine, an instrument that lets us see into what Shakespeare's Prospero calls "the dark backward and abyss of time." Prospero's daughter, Miranda, professes not to be curious about her past: "More to know did never meddle with my thoughts," she says. But as a species we are infinitely curious about our past, which is why we invest billions of dollars of our collective wealth in a magnificent instrument that orbits 350 miles above the surface of the Earth and peers deeply into time. That dot of light observed by the Hubble at the beginning of material creation is a faint glimpse of the cosmic forge in which our very atoms were created.

The most reliable meteor shower of the year is the **Perseids** of August. Anytime during mid-August is a good time to look for "shooting stars," but the nights of August 11–12 are best, especially after midnight. You might see streaks of light in any part of the sky, but trace back their paths and most of them (the Perseids) will appear to radiate from a place in the constellation Perseus, called the **radiant** of the shower. In mid-August, the Earth in its orbit around the Sun passes through the track of a comet, a part of space littered with dust shed by the comet during its many previous passages through the inner solar system. When one of these bits of matter—typically the size of a sand grain—collides with the Earth's atmosphere, it is heated by friction and vaporized. We see the streak of burning vapor as a "shooting star" or "falling star," not a star at all but a smidgen of comet dust swept up by our planet. (Other good times of the year for meteor watching are listed in appendix 2.)

Not far from the Perseid radiant is the famous **Double Cluster of Perseus**, an exciting object through binoculars or a small telescope. But optical aid is not essential; wait for a perfect Moonless night and look for a blur midway between the vee of Perseus and the less bright end star of the em of Cassiopeia. The two young clusters stand side by side, like newborn twins. Each contains thousands of stars, condensed by gravity from their parent nebulae of gas and dust within the last few millions of years. Curiously, the Double Cluster of Perseus did not make it into Messier's catalog of fuzzy spots.

A fuzzy spot that did make it into the catalog is **M31** in Andromeda. You will need a very clear, very dark night to see it, but it is well worth looking for. It is the most distant object you will ever see with your naked eye. This is the **Andromeda Galaxy**, a gorgeous spiral companion to the Milky Way, containing one trillion stars. What you see with the unaided eye is just the bright central part of the spiral. The light that enters your eyes left the Andromeda Galaxy 2 million years ago—at a time when our earliest human ancestors were scanning the night sky with the first dim stirrings of consciousness. Look for a faint blur of light near the second star out along Andromeda's body, on the fainter, concave side of the vee.

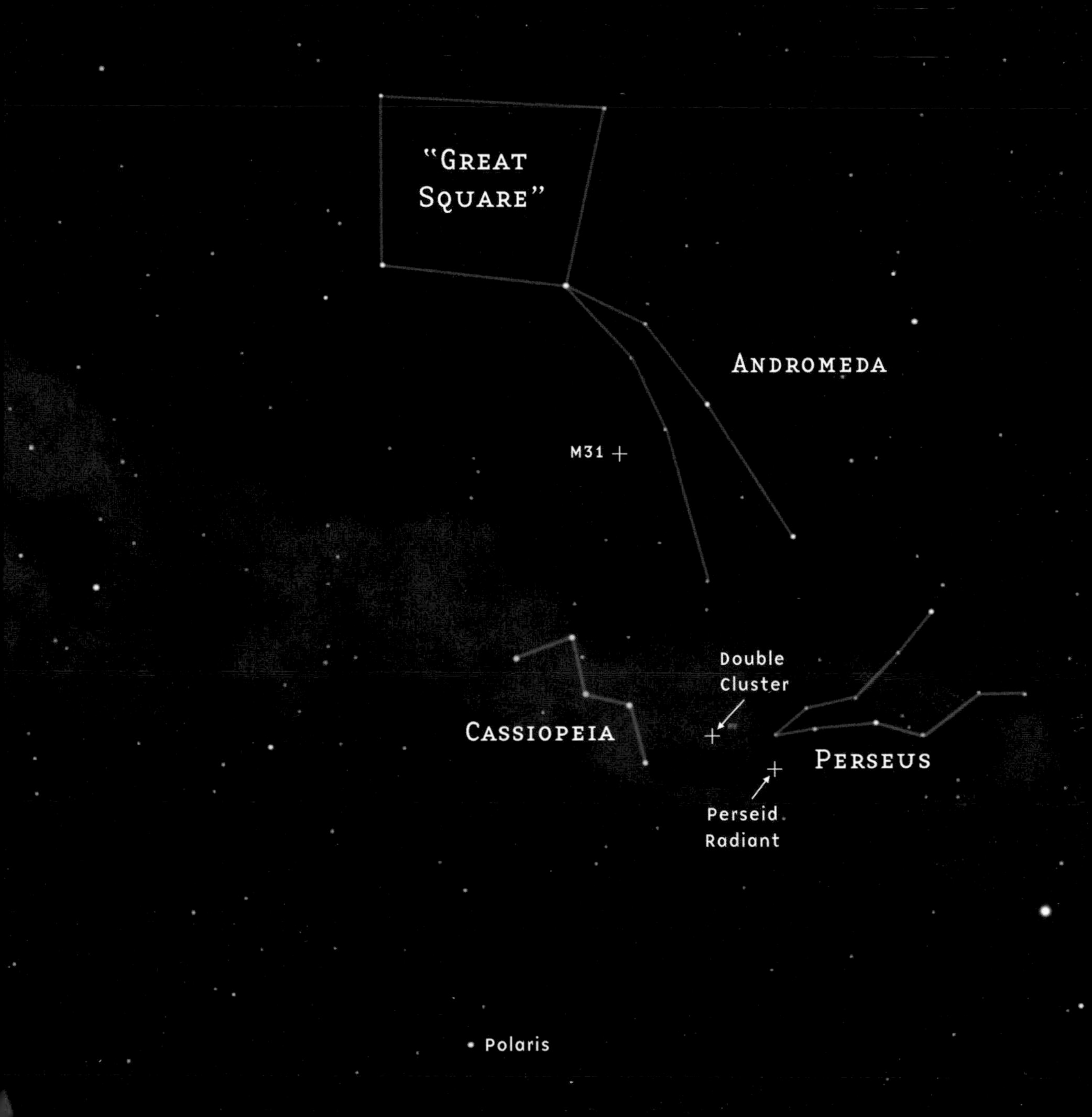

"GREAT SQUARE"

ANDROMEDA

M31

Double Cluster

CASSIOPEIA

PERSEUS

Perseid Radiant

Polaris

Perhaps nothing you will see in the sky with your unaided eye so rewards the application of a vigorous imagination as M31 in Andromeda. To the eye, a faint blur, a smudge on the dark windowpane of night. To the imagination, an island universe, a whirling pinwheel of one trillion suns, the magnificent companion spiral to our own Milky Way Galaxy.

With the map as your guide, let your imagination supply the third dimension to space. The disk of our own galaxy crosses the sky from Deneb to Capella. We see it as a pale band of light. We are inside the disk. All of the stars you see on the map are strewn around us in our own neighborhood of the Galaxy. We look somewhat up out of the plane of the Galaxy and see another spiral, 2 million light-years away. What lies between our Milky Way Galaxy and the Andromeda spiral? Perhaps nothing at all. Certainly nothing luminous. Nothing that absorbs or radiates light. It is within the galaxies that the stars, dust, and gas of the universe are concentrated. It is within galaxies that stars are born and die.

The Andromeda Galaxy is the nearest spiral galaxy to our own, part of the **Local Group** of galaxies, most of which are raggedy dwarfs. Observatory photographs show the Andromeda Galaxy as a splendid copy of our own galaxy, seen nearly edge on, but tipped just enough to give a hint of spiral arms. If your imagination has interpreted the map correctly, you will understand that the view of the Milky Way Galaxy from Andromeda would be similar—a spiral tipped at a slight angle.

As discussed earlier, the Andromeda Galaxy is quite a large object in the sky, about four times broader than the full Moon. You could just cover it with your thumb at arm's length. We see only the bright central nucleus of the galaxy with the unaided eye. How many stars are in the Andromeda Galaxy? In my astronomy classes I make a model of a spiral galaxy on the floor by pouring out a box of salt. To have as many grains of salt as there are stars in a great spiral—such as the Andromeda or Milky Way Galaxies—would require 10,000 one-pound boxes of salt!

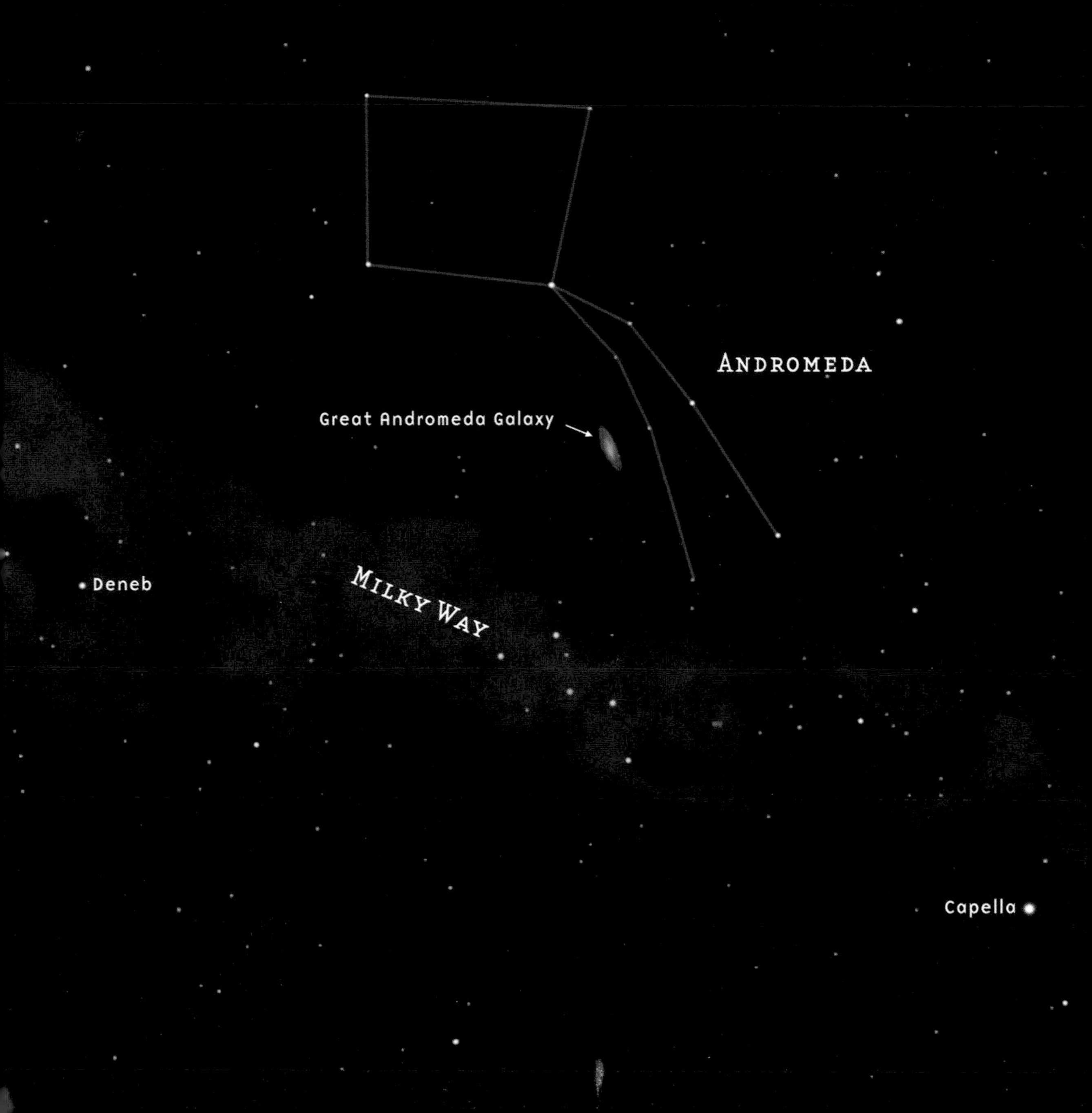

ANDROMEDA

Great Andromeda Galaxy

Deneb

MILKY WAY

Capella

11. Spirit

Life in the Universe

The hunt is on. The scent of the prey is in the air. The hunters are working themselves into a frenzy of excitement. The quarry? Dark matter. The unseen substance that seemingly constitutes as much as 90 percent of the universe, the hidden mass that shaped the universe's beginning and (as we will see) will determine the universe's ultimate fate. There is glory to be garnered, prizes to be won, by the astronomers who first identify what it is. Imagine! An all-pervasive dark continent of matter, constituting most of what exists in the cosmos, waiting to be discovered. The ultimate *terra incognita*. The Fountain of Youth, the Seven Cities of Cibola, and the Source of the Nile—all rolled into one.

The existence of this hidden stuff was first guessed by astronomer Fritz Zwicky in the early 1930s. He measured the relative velocities of galaxies in a cluster of galaxies in the constellation Coma, and he estimated the total mass of the galaxies by adding up all of their luminous matter. The problem: The galax-

Astronauts Story Musgrave and Jeffrey Hoffman service the Hubble Space Telescope during the 1993 mission to repair the telescope's faulty vision.

ies are moving so fast relative to one another that they should long ago have escaped the gravitational attraction of their companions. In the time since the galaxies formed, more than 10 billion years ago, the members of the cluster should have dispersed. Yet the cluster endures. Something nonluminous is holding it together. Something with mass. Something with lots of mass.

In recent years, more and more evidence has been added to Zwicky's original observations to suggest that galaxies and clumps of galaxies contain vastly more matter than is apparent from the light of stars. What is this missing stuff? Around our corner of the cosmos—the solar system—most of the known matter is concentrated in our star, the Sun; everything else (planets, moons, asteroids, comets, meteoric dust) is just a drop in the bucket. Whatever the hidden matter is, it is certainly not the kind of familiar objects that populate our immediate neighborhood. A number of candidates have been suggested for dark matter. Among the most eagerly sought: MACHOs (massive compact halo objects), huge numbers of planetlike objects too small to have ignited as stars, or burned-out cores of stars, forming a vast dark halo around galaxies; and WIMPs (weakly interacting massive particles), subatomic particles filling the universe, that were created in myriad numbers in the big bang, but which are almost impossible to detect because they interact so weakly with ordinary matter.

And so the search is on. Armed with telescopes, satellites, computers, and exquisitely sensitive particle detectors, a slew of clever astronomers and physicists are hoping to become the Christopher Columbus of dark matter. If dark matter can be proven to exist, in the amount predicted, then we will gain confidence in our understanding of the universe. If dark matter remains elusive, then our present understanding of nature may be seriously flawed. The riddle is well worth solving; the search is of profound importance.

But beware of hype. You will hear dark matter called "the Holy Grail of cosmology." Other astronomers have proposed that the detection of MACHOs or WIMPs would be the "ultimate Copernican Revolution." Copernicus displaced us from the center of the cosmos; confirmation of dark matter will bump us even farther from prominence. "How much farther can we fall?" asked one astrophysicist. If dark matter exists in the quantity sought, we are "an insignificant bit of 'noise'—a cosmic afterthought," he said. *Whoa!* Not so fast. The mere fact that most of what has weight presently eludes our senses hardly constitutes a revolution on the same scale as that of Copernicus.

Most people have long since accepted our spatial and temporal insignificance in the universe. Whether life is a cosmic afterthought or a cosmic forethought remains an open question, but detection of dark matter has nothing to do with how we value ourselves. We've known for a long time that we are materially an insignificant fraction of the universe. The entire biomass of Earth is negligible compared to the mass of the planet, and the mass of the planet is negligible compared to the mass of the Sun. If all luminous matter in the universe turns out to be only the tip of the gravitational iceberg, that will hardly change the equation. In that sense, at least, we have no farther to fall.

It is not the amount of stuff that matters, but the *complexity* of stuff. Visible matter (stars, planets, and glowing gas) may be only 5 or 10 percent of the universe, and living matter may be only an infinitesimal part of the visible universe, but so far human beings—yes, you and me—are the most complex things we know about in the universe. An amoeba is vastly more complex than a star. A mushroom is more complex than an entire universe of dark matter. Far more significant for our sense of cosmic importance would be detection of an intelligent signal from space, from some other island of intelligent complexity. Knowing with certainty that the flame of intelligence burns brightly among the stars would indeed be a revolution to rival that of Copernicus.

If memory serves me right, my first true love was Princess Aura, daughter of Ming the Merciless of the planet Mongo. Aura had a thing for Flash Gordon, and pointy-nosed Ming was smitten with Flash's girl companion, Dale. I can't remember how these attractions worked out, but we can be sure they had chaste conclusions after some titillating preliminaries. It was from Flash's adventures on Mongo that I learned the First Law of Alien Life: All women on other planets are young, beautiful, and scantily clad, and all men are beastly, misshapen, or otherwise unattractive. The First Law had a corollary that gave hope to all of us mostly male sci-fi fans: If we ever make contact with extraterrestrials, even a halfway decent-looking human male will be much in demand. No wonder, then, that male astronomers of Flash Gordon's generation have invested so much interest in the search for alien life. If you've grown up with Princess Aura, Queen Undina (of Mongo's undersea kingdom), and Queen Fria (of Mongo's ice kingdom), all utterly alien and utterly beautiful, then—well, then other worlds beyond the Moon start looking pretty good. This, of course, was before the days when terrestrial women joined terrestrial men as equals on space missions, and—to a far greater extent than yesteryear—in observatories. Gender equality on Earth has undergone a Copernican Revolution of sorts. It remains to be seen whether the First Law of Alien Life still applies in other planetary systems.

Life may exist elsewhere in our own solar system, but no one realistically expects to find intelligent life on Mars or the moons of Saturn. If we are going to make contact with a Ming the Merciless or a Princess Aura, we must look to other stars. We have just entered the era when detection of planets around other nearby stars is possible, though not directly. Telescopes do not yet have the resolving power to separate the faint reflected light of a planet from the overwhelming glare of its parent star, although this may soon change. So far, several dozen planets have been detected orbiting nearby stars, discerned either by their gravitational effect on the star (they cause the star to wobble as they go around) or by periodically blocking some of the star's light (if the orbit of the planet is lined up just right with Earth). Both ways of detecting planets favor big Jupiter-like planets close to their parent star, and these are exactly the sorts of planets that have been discovered. For example, the first of the newly discovered planets—orbiting the Sun-like star 51 Pegasi—has a mass 150 times that of Earth and circles its star even closer than Mercury orbits the Sun.

Unfortunately, the discovery of such planets does not do much to keep our Flash Gordon dreams alive. If we assume that the planet's average density is about that of Earth, creatures on such a massive planet would weigh five times more than on Earth, a crushing load. The planet might support ground-hugging centipedes, maybe, but no tall, willowy princesses. And then there's the problem of the planet's proximity to its star. The surface temperature would be in excess of one thousand degrees Celsius. No water for Queen Undina, no ice for Queen Fria. Nothing but scorched rock (if that's what the planet's made of, an unlikely possibility).

Another newly discovered planet orbiting the star 47 Ursae Majoris might seem to offer a better hunting ground for aliens. That planet is about twice Earth's distance from its star, which suggests more tolerable temperatures. However, the planet is even more massive than the one near 51 Pegasi, with a surface gravity eight times greater than on Earth (assuming an Earth-like density). Even ground-hugging centipedes would feel the burden.

This brings us to the Second Law of Alien Life that we learned from Flash Gordon: The dominant creatures on other planets will always be at a stage of evolution approximately the same as our own. So Flash Gordon, for example, will be on the same psychosexual wavelength as Queen Azura of Mongo, from whom Earthlings acquired the idea of the string bikini. And in this respect, the new planetary discoveries hold some promise. The lifetime of a star is determined by its mass. Stars much more massive than the Sun burn out quickly, not allowing time for life to evolve to our level of physical and intellectual development. Stars much less massive than the Sun live virtually forever. Some have been around since the beginning of the Galaxy, time enough for evolution to far surpass our level of development. Most of the newly discovered planets orbit stars roughly similar to our own—yellow, midsize, midmass stars. These stars live long enough for life to reach and perhaps surpass our level of development, in conformity with the Second Law of Alien Life, but not so long as to leave us far behind.

And while we're at it, the Third Law of Alien Life is also worth considering: All intelligent extraterrestrials speak English. This makes it possible for Ming the Merciless to say things like this to lovely Dale, which stands as one of the least successful pick-up lines in history: "The reason for our success is that we possess none of the human traits of kindness, mercy or pity! We are coldly scientific and ruthless! You'll be one of us." The subtle psychological implications of the Third Law work to the advantage of the astronomers who are making these important new extrasolar planetary discoveries. Their research requires generous government funding, and politicians are more likely to favor the search for extrasolar planets and alien life if they can reasonably expect to understand what the aliens have to say.

Meanwhile, the discovery of supermassive or sizzling-hot planets in nearby solar systems is interesting, but we are all waiting for a Mongo—a real, honestto-goodness Earth-like planet at a comfortable distance from a Sun-like star. When that happens, the politicians will really sit up and take notice. There's lots of old Flash Gordon fans out there who have not forgotten their early infatuations with Princess Aura.

But more seriously, my whimsical Second and Third Laws of Alien Life are in fact relevant to the search for extraterrestrial intelligence. If life in another star system has not reached at least our level of technological development, we will never detect it, because we are not going to travel between the stars, at least not with any currently imaginable technology. Contact can only be made by detection here on Earth of an electromagnetic signal—most likely radio waves—of an obviously intelligent origin. On the other hand, if an extrasolar civilization is too much in advance of our own, it may have discovered ways of communicating that leave us entirely in the dark. And if we do make radio contact by detecting a signal, the "message" will certainly *not* be in English—or Chinese, or any other human language—and it might be undecipherable. We have only one example of intelligence to guide our efforts at decoding an extrasolar signal, and the source civilization may be so utterly alien to our own that we will never understand what it is saying.

But wait. Why is it impossible that we will go there, or they come here? With face-to-face contact, no decoding will be necessary. The existence of ETs will be a matter of direct observation. Ah, yes. Galactic travel. A zip through hyperspace, the meat and potatoes of science fiction. "Shall we go to warp speed, Captain?" We have traveled so often to Antares and beyond in books, films, and television, that it seems only a matter of time before humans must actually embark upon such a voyage. *Impossible*, says the physicist. *No way*, says the astronomer. To get to the nearest star using anything like present technology would be a journey of tens of thousands of years. At the speed of light, a trip to Alpha Centauri could be made in a very reasonable 4.3 years, but an infinite amount of energy would be required to accelerate a spaceship to the speed of light. As Sir Richard Woolley, once Britain's Astronomer Royal, so tidily put it: "Space travel is bilge." But don't tell that to the students in my astronomy class. They listen politely to what I have to say about the unbridgeable distances to the stars, about the energy requirements for travel at speeds close to the speed of light, and about nature's prohibition on travel at speeds greater than the speed of light; they listen politely, and then they say: "But how can you be sure? Maybe there's lots of stuff we don't know yet." And, of course, they are right. In 1957, Dr. Lee De Forest, the man who invented the electronic vacuum tube, said that humans would never reach the Moon "regardless of all future scientific advances"—and we know how quickly *that* prediction went awry. One hundred years ago, who would have imagined crossing the Atlantic or Pacific Oceans at twice the speed of sound? Humans will almost certainly be colonizing Mars before the middle of this century. Can Antares be far behind?

Science fiction writers offer lots of ways to get to Antares fast. There's old reliable "hyperspace." According to this view, our familiar three-dimensional universe is folded or crumpled in a higher dimension, as a flat two-dimensional piece of paper might be crumpled in three-dimensional space. The space traveler then moves quickly from one place in the universe to another by taking a shortcut through the higher dimension. A variation on this theme imagines a shortcut through a black hole, which in the mathematics of general relativity can actually be a sort of hyperdimensional tunnel connecting two widely separated parts of the universe (a so-called wormhole). Or one might avoid nature's prohibition on faster-than-light space travel by taking a quantum jump. In the standard interpretation of quantum physics, electrons in the atom jump from one energy level to another without traversing the space (or energies) in between. If electrons can do it, why not the spaceship *Antares Bound?* And if this stretched-out physics doesn't do the trick, one can always explore psychic dimensions. Dematerialization. Travel from one place to another as pure thought. Reconstitute the spaceship as bodiless information, instantly communicated within that great Mind which is the universe.

Why not, indeed? But for the moment it's best to stick with what we know to be possible, and what we *think* to be impossible. To imagine that *anything* is possible—as do the wildest of the science fiction writers—takes the edge off our current aspirations. To imagine that the *impossible* is carved in stone takes the edge off, too. Science is different from science fiction because it cleaves to the boundary between the possible and impossible. It is a delicate balancing act—to be skeptical and open at the same time. Given our present knowledge, space travel to other stars is indeed "bilge," and we should invest our best efforts in other ways of communicating with extraterrestrials. But we also know that our present knowledge of the world is partial and tentative. Someday, humans may travel to Antares. Someday Antareans may visit Earth. We should be prepared to be astonished.

In his autobiography, the brilliant physicist John Archibald Wheeler makes this confession of faith: "Whatever can be, is." He goes even farther. "Whatever can be, must be," he says. What he means is this: Anything that is not prohibited by the laws of nature exists. For example, the theory of general relativity allows for the existence of wormholes, those curious kinks in the fabric of space-time that connect remote places: Fall into a wormhole at one place in the universe, pop out somewhere else, perhaps trillions of miles away. Of course, no living organism could survive such a journey, but Wheeler confidently asserts, "If relativity is correct, and if it allows for wormholes, then somewhere, somehow, wormholes must exist."

Whatever can be, must be. A simple philosophy to be considered skeptically, but one with profound human implications if true. It means, for instance, that we are not the be-all-and-end-all of creation, because certainly we are not the most complex or intelligent life-forms consistent with the laws of nature. If the laws of physics and chemistry do not prohibit creatures more intelligent than ourselves, then according to Wheeler's principle such creatures must exist. This is why scientists are busily searching for intelligent life in the universe. Researchers on the SETI project—Search for Extraterrestrial Intelligence—are scanning the heavens for intelligent radio signals from space, examining the radiation from thousands of stars with powerful computers, looking for any pattern that could be interpreted as the product of intelligent beings.

These researchers are currently planning the world's biggest radio telescope dedicated to the search for intelligent signals from space. The telescope will consist of 500 to 1,000 small dishes with a combined area of one hectare (about 2.5 acres). The dishes will scan the sky in unison. The telescope will increase by a factor of ten the efficiency of the present search for intelligent life. Tens of thousands of Sun-like stars will be monitored. If anyone is beaming a recognizable signal our way, the new scope will have a good crack at finding it. The One Hectare Telescope is conceived as a precursor to a Square Kilometer Array, a radio telescope one hundred hectares in combined area that astronomers want to build by the year 2010. The Square Kilometer Array will be powerful enough to pick up radio "leakage" from nearby planet systems. Members of a technological civilization need not be beaming a signal our way to be detected; they need only be using radio waves to communicate among themselves.

These efforts will ultimately cost hundreds of millions of dollars. Is the investment worth it? If another intelligent civilization is detected, it will be the biggest news in history and the best money ever spent. And if Wheeler's principle is correct, that civilization is out there. But is Wheeler right? Only time will tell. In the meantime, astronomers will continue looking for life within our own solar system—on Mars, on comets, in the ice-covered ocean of Jupiter's moon Europa. The discovery of even single-celled organisms on another celestial body would make the pervasiveness of life in the universe seem much more likely.

Some scientists seriously consider the theory known as *panspermia*, which holds that seeds of life are adrift throughout the Galaxy, now and then alighting upon a favorable planet where they multiply and flourish. By this account, all terrestrial life—including you and me—may have descended from an intergalactic drifter. On the other hand, if life arose independently on Earth, without precursors, from the amalgamation of inanimate materials, it still doesn't rule out the possibility that animation (or intelligence) is pervasive. The chemistry of carbon-based terrestrial life presumably applies throughout the universe, and the universe presents us with the prospect of hundreds of billions of galaxies, all replete with stars and planets. The number of worlds, and therefore habitats, is unimaginably large, perhaps infinite. Who is willing to bet against any possibility in all that vastness? Whatever can exist, must exist, guesses Wheeler, and since our finite minds have only the dimmest notion of nature's rules of possibility, we can hardly rule out anything. If Wheeler's conjecture is true, of this we can be sure: We are not alone. Somewhere among the multitudinous galaxies are creatures both more and less intelligent than ourselves.

Now we have come full circle, around the ecliptic, back to the place where we started in Taurus, the Bull. Face east this evening, and watch Taurus rise, preceded by the Pleiades, a first sign of winter. Also rising before Taurus are the last two signs on our zodiac tour, Pisces and Aries. We are still traveling through an empty quarter of the sky, looking down out of the Milky Way Galaxy. Pisces will be hard to find under the best of conditions. Aries has two stars bright enough to attract only modest attention.

Pisces *(PYE-sees),* the Fish, is usually shown on ancient star maps as two fish bound together by their tails with a cord looped around the star Al Rischa. The star's name means simply "the cord." In Greek mythology, when the gods were pursued to Egypt by the monster Typhoon, Aphrodite, the goddess of love, and Eros, her son, disguised themselves as fish and hid in the Nile. Now we find them swimming in the sky. How their tails got tied together is anybody's guess.

Aries *(AY-rih-eez),* the Ram, was the disguise adopted by Zeus in his flight from Typhoon. Another story of the Ram centers on Phrixus, the son of Athamus, king of Boeotia. A nasty rivalry between the mother and stepmother of Phrixus led to a drought and famine in Boeotia. Athamus was given a false report, presumably from the Oracle of Delphi, that he must sacrifice his son to restore the fertility of the land. This he was about to do when the god Hermes sent a ram with a golden fleece, who carried away Phrixus and his sister, Helle. As the magical ram flew over the narrow body of water connecting the Black and Aegean Seas, Helle fell off and drowned; that strait is known to this day as the Hellespont. In thanksgiving to Zeus for his rescue, Phrixus sacrificed the ram (a curiously unappreciative act, from the ram's point of view) and gave its golden fleece to Aeëtes, the king who offered him refuge. Aeëtes nailed the golden fleece to a tree in the sacred grove of Ares, where it was later snatched by Jason and the Argonauts.

Here, too, near Pisces and Aries is the missing character of the story of Andromeda and her rescue by Perseus—Cetus, the Whale, an inconspicuous constellation.

Pisces
Aries
Ecliptic
Al Rischa
Cetus
Pleiades
Taurus
Looking East

On star maps and globes, lines like the ecliptic and the celestial equator are marked for all to see. As we stand under the night sky, we must use our imaginations to "see" these lines in the sky. The place among the stars where the equator crosses the ecliptic has had a special significance in astronomy since the earliest days of sky watching. This place is called the **vernal equinox,** or spring equinox, and it is where we see the Sun as it crosses the equator back into northern skies on or about March 21. It is called "equinox" because on this day we have equal hours of daylight and darkness (*nox* = night). In modern astronomy, this is the place among the stars from which we measure east-west positions of celestial objects—the equivalent of Greenwich, England, on terrestrial maps. The equinoxes and solstices refer to both *places* on the celestial globe and *times* when the Sun is there.

Because of the precession (wobble) of the Earth's axis, the equinox slowly drifts westward along the ecliptic, moving through one zodiac constellation every few thousand years. Thousands of years ago, when astronomers first worked out these things, the vernal equinox was in Aries, and modern astronomers still refer to the place of the equinox as "the first point of Aries," even though it now lies in Pisces. Astrologers, too, still work out their horoscopes as if the Sun were in Aries on the first day of spring. Six hundred years from now, the vernal equinox will cross the border between Pisces and Aquarius. This will be the "Dawning of the Age of Aquarius" made famous in a lyric of the rock musical *Hair.*

The unfolding of our knowledge of the universe has been a long tension between what we see and what we imagine. Astronomy is an art of the imagination: imagining dots of light as Sun-like globes of blazing gas; imagining "fuzzy spots" as star clusters, exploding stars, and galaxies; imagining a web of lines and numbers flung across the dark globe of night; imagining a third dimension—depth—where unaided vision presents only two. Always we are trying to tune the tension between what we see and what we imagine, testing the one against the other, seeking a true picture of this amazing universe we inhabit. *We are truth-seeking pilgrims among the stars.*

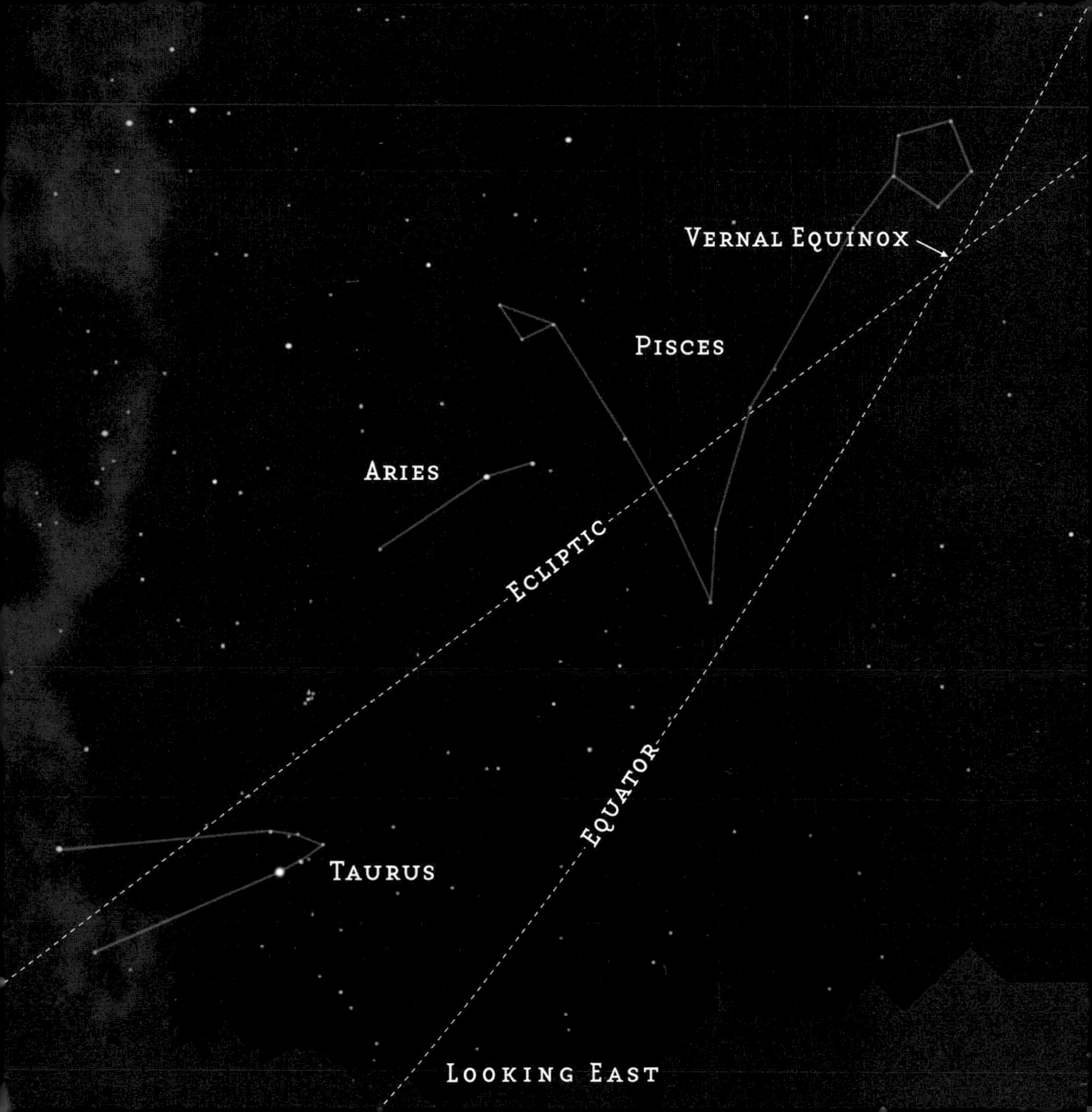
Vernal Equinox
Pisces
Aries
Ecliptic
Equator
Taurus
Looking East

12. Omega

How the Universe Will End

Imagine the following experiment. Remove all visible life from planet Earth. Get rid of the elephants, tigers, apes, dogs, cats, and fleas. Birds, fish, worms, and beetles. Humans too. Remove the plants. Trees, flowers, seaweed, grass. The night birds and insects that sing while we watch the stars. The blossoms that scent the night air. Everything. Eliminate, too, all the invisible microbes in air, soil, and sea. Then, when not an iota of living matter appears to remain, zap the surface of the Earth with a killer blast of one-thousand degree heat to kill off any organisms that managed to escape our attention. Finally, just to be sure, boil away the sea. Scrape off every grain of loose soil down to bedrock, every particle of sediment from the ocean floor, and haul it away to some other planet. Scrub the rock bare. The Earth is now sterile. Life is finished. The whole glorious terrestrial experiment with animation and spirit has come to an end.

Right? Wrong.

Two spiral galaxies collide. Eventually, billions of years from now, they will merge into a single massive galaxy.

Earth would still harbor an amount of life perhaps as great in mass—if not as diverse—as what we got rid of. Where? In microscopic pores and fissures of the rocky crust, extending several miles (at least) below the surface. Viable bacteria have been found a mile below the surface of Washington State in bare volcanic rock. They have been recovered from two miles down in ancient sedimentary rocks of Virginia, where they have lived out of touch with surface life since the time of the dinosaurs. Scientists now realize that *pound for pound* there may be just as much life below the surface of the Earth as above. The crust of our planet teems with living organisms.

Surprised? So are scientists.

Solid rock may not seem a likely environment for life. But the crust of the Earth is full of microscopic pores and fissures, through which water percolates. All rocks within a mile or so of the surface are saturated with water. And the water contains bacteria. How do the microorganisms survive in total darkness, cut off totally from the atmosphere and sun? By living off the internal heat of the planet. Subsurface rock is hot; the deeper you go, the hotter it gets. Deep subsurface bacteria take in water and carbon dioxide and use thermal energy to metabolize carbohydrates, releasing methane and hydrogen sulfide waste. These organisms may even live off the rock itself, rock that is "weathered" by the water perking through it, releasing useful hydrogen. It's a grim sort of life, a kind of permanent hell. Deep subsurface bacteria must survive high temperatures that would kill more familiar forms of life. They may reproduce only once a year, or even once a century, compared to the minutes or hours that are typical of their surface cousins. Deep bacteria are about as close to being dead as something can be and still be alive. But alive they are, living out their lives at a languorous geologic pace. Brought to the surface, many subsurface bacteria can be cultured in a laboratory. They are undoubtedly related by common descent to their surface cousins.

We have much to learn about life inside the Earth. Scientists have literally only scratched the surface. We need to drill more deep holes in a variety of environments on continents and on the sea floor. We need to bring up samples of deep subsurface rock with particular care to avoid contamination by surface organisms. How much life is down there? How does it live? What is its role in creating petroleum and mineral deposits? What role does it play in the overall ecobalance of the Earth? Most interesting of all: How is subsurface life related to life on the surface? Did subsurface bacteria migrate there from the surface, carried deep by percolating water or buried with surface sediments that became sedimentary rock? If so, then deep subsurface bacteria have been out of touch with their surface cousins for tens of thousands or even hundreds of millions of years. Or did life begin in the depths and only later make its way to the surface? If so, then even you and I may be descended from bacterial troglodytes.

These are big questions, important questions, and they have the potential to change our understanding of life in the greater universe. If life can survive in the hellish conditions that exist miles below the Earth's surface, with a kind of slowed-down metabolism, then the odds dramatically improve for finding life in such apparently inhospitable places as the Moon, Mars, or the frozen seas of Jupiter's moon, Europa. It also becomes less difficult to imagine that seeds of life may be adrift in interstellar space or carried from place to place by meteorites or comets. Perhaps Earth's subsurface microbes have in times past been carried to Mars or Venus as passengers in rocks blasted from our planet by asteroid collisions. Or perhaps it was the other way around; perhaps life on Earth came here as passengers on rocks blasted from the surface of Mars. (Meteorites of Martian origin have been found on the Antarctic icecap, and at least one of them bears tantalizing hints of biological activity, although this interpretation of the evidence is highly controversial. We know the meteorites are from Mars because trapped bubbles of gas in the meteorites exactly match the composition of the Martian atmosphere sampled by the Viking landers.) This much is certain: Life can survive under conditions of elevated temperatures and nutrient impoverishment that previously were thought to be impossibly hostile. The discovery of deep subsurface terrestrial bacteria stretches our perception of the possible.

It also says something about the permanence and resiliency of life. Even if the Earth's surface were utterly sterilized by a catastrophic asteroid impact, the explosion of a nearby supernova, or a full-scale nuclear war, the planet would go on living. Microbial gnomes that live in the pores of deep rock would migrate to the surface, adapt to air and sunlight, and start all over again. Life, it would seem, is in it for the long haul; its tenacity is unbounded. Comets will collide with planets. Stars will come and go. Somehow, somewhere, life will go on. Its fate is ultimately tied to the fate of the universe itself.

Scientists are pretty sure how the universe began. They are less certain how it will end.

The beginning was explosive. The universe had its origin about 15 billion years ago in an expanding fireball of radiant energy. Space and time unfurled from a tiny seed of infinite energy, like a balloon inflating from nothing, cooling as it swelled. Energy became matter; matter became stars and galaxies, racing outward. Today the galaxies continue to fly apart, impelled by their initial impetus, bearing clusters of galaxies to ever greater separations.

What is less clear is where the universe will go from here. There are two possibilities: Either the universe will expand forever, carrying the galaxies always farther apart, into cold and darkness, infinitely dispersed; or it will cease expanding and begin to contract, the galaxies drawing closer and closer, ending as it began in a blaze of radiant energy. A whimper or a bang? A terminal glide into Stygian gloom, or annihilation in blinding light? The situation is rather like shooting a projectile up into the air: Will it climb forever, escaping the gravitational pull of the Earth? Or will it slow to an instantaneous stop, then fall back to Earth, gathering speed? It all depends on whether the initial velocity of the projectile imparts sufficient energy to overcome Earth's gravity, the so-called escape velocity. At the Earth's surface, the escape velocity is 25,000 miles per hour, which is why it takes a pretty big rocket to launch a spaceship to the Moon or planets. Like a rocket engine, the big bang hurled the galaxies outward; gravity is pulling them back together. Do the galaxies have sufficient speed to overcome their mutual attraction? Do they possess the escape velocity? The question turns out to be surprisingly difficult to answer, not least because of dark matter, that unknown, nonluminous *stuff* (whatever it is) that might constitute as much as 90 percent of the mass of the universe.

Two things are important to know: What is the present expansion rate of the galaxies? And how much matter is acting to slow down the expansion? During recent years astronomers have made tantalizing progress toward finding answers to both questions. Several groups of researchers have been studying supernovas in distant galaxies. These extremely bright exploding stars, which can be observed billions of light-years away, are used as indicators of the universe's changing expansion rate. The recessional velocities of the supernovas (and the galaxies they are in) can be deduced from a stretching of the wavelengths of their light, just as the pitch of an automobile's sound is lowered as it races away from us. The data indicate that the expansion of the universe is not slowing down enough to make the galaxies fall back upon themselves. It may even be possible that some yet unknown force is *accelerating* the expansion, driving the universe inevitably toward infinite dispersal. Other groups of astronomers have been comparing the actual distribution of galaxies to computer models for how the universe should evolve with different densities of matter. The best current fit between observation and calculation assumes that there is not enough matter to stop the expansion. And finally, observations of the so-called cosmic microwave background radiation—the light from the big bang stretched and cooled by the universe's expansion—would appear to confirm that the universe will likely expand forever.

So what is the best present guess for how the universe will end?

Five billion years from now the Sun will swell to become a red giant star. Its surface will balloon outward toward the Earth, cooling and reddening (even as the core collapses and heats up). Mercury and Venus will be consumed, and a bloated red star will fill Earth's sky. Of course, all life on the Earth's surface will have long since been extinguished; as soon as the Sun starts to swell, the atmosphere and oceans will be boiled away and the surface sterilized. Perhaps microbes in the deep rock will survive for a while, but ultimately the rocks themselves will melt. It is possible that the Sun will puff off enough of its mass so that its gravitational pull will diminish and Earth will drift outward, staying outside of the Sun's swelling envelope. Or maybe tidal forces between the Earth and Sun will slow the planet down, and it will be drawn inexorably into the Sun's depths. Life on Earth will be extinguished either way. After some hundreds of millions of years as a red giant, the Sun will collapse to a glowing ember—a white dwarf—that will slowly fade from sight. If the Earth has survived being consumed by the star, a cold, lifeless planet will circle the tiny darkening Sun. But who can predict the destiny of terrestrial life? Will the doomed planet have thrown off spores into interstellar space? Will our human (or posthuman) descendants have discovered ways to travel among the stars? Will Earthlings have been incorporated into a greater galactic civilization? Your guess is as good as anyone's.

After the Sun's death, 100 billion years will pass and the universe will be stretched exceedingly thin. Local clusters of galaxies will have amalgamated into supergalaxies, and supergalaxies will have drifted far apart. Within the galaxies, the last dregs of energy will be squeezed out of star-birthing nebulae. No new stars will be born. The sky will grow increasingly dark. All ordinary matter will be compacted in dead stars—the cold cinders of white dwarfs, and those denser stellar remnants called neutron stars and black holes—or cold lumps of rock (perhaps even ghostly, lifeless spaceships) adrift in the darkness. Life, which requires the extraction of energy from its environment to survive, will be increasingly hard-pressed for resources. Planet by planet, the last flickers of animation and spirit will be snuffed out. Somewhere, in a last pool of cosmic warmth, perhaps in a faraway galaxy, a final organism will expire. Life, which for billions of years had burned among the stars like a cool blue flame, will flicker out. A dead universe will slide into black oblivion.

Each of us will react to this doomsday story in our own way. When I was younger, I was drawn to a universe that ended as it began, in a blazing fireball of energy, perhaps to bounce back and start all over again, a new big bang. But now, as I slip past middle age, I find something grandly sedate and dignified about the new predictions for the universe's demise, something that suits my mature mood. No recurring fireballs for me, please. I'm ready for that long, dark cosmic nap.

As the year ends, we return to where we began, to the stars of winter ablaze in the night: Orion, Taurus, Sirius, and the Pleiades. The Sun hangs low in the daylight sky. This is the time of year our ancestors celebrated the Sun's rebirth, its turning from its southern journey back toward northern skies. Many of the world's religions celebrate feasts of light at this time of year (Christmas and Hanukkah, for example, in the Judeo-Christian tradition). It seems an appropriate time to consider the mystery that manifests itself in the night sky—that infinite well of darkness containing galaxies, stars, planets, and perhaps pervasively the stuff of life and spirit. This is the week at the end of the year when wide-eyed children everywhere will open packages, when mysteries wrapped in colored paper and ribbons will be unraveled. Einstein, whose theory of general relativity guides all of our speculations about the universe's fate, once said that science is a sphere of activity in which we are permitted to remain children all of our lives. "What I mean," he explained, "is that we never cease to stand like curious children before the great Mystery into which we are born."

A common misconception about science is that it's somehow inimical to mystery, that it grows at the expense of mystery and intrudes with its brash certainties upon our sense of wonder. Many times students have complained to me that science "takes the mystery out of the world." In reply, I offer the metaphor of *knowledge as an island in a sea of infinite mystery* and point out that the finite extension of our knowledge hardly depletes the sea. Rather, the growth of knowledge extends the shore along which we might encounter the thing that Einstein sometimes spelled with a capital *M*. To Queen Elizabeth of Belgium, he wrote: "It gives me great pleasure to tell you about the mysteries with which physics confronts us. As a human being, one has been endowed with just enough intelligence to be able to see clearly how utterly inadequate that intelligence is when confronted with what exists." This was profound humility from a man who spent his entire life using his intellect to extend our knowledge of the universe.

Einstein had no use for those people who sought mystery in paranormal fads and superstitions. Nor did his deeply religious nature lead him toward any sort of God fashioned in the image of man. His religion was "humility" in the face of the magnificent structure of nature that can only be imperfectly comprehended. Christmas and Hanukkah celebrate light that comes into darkness and illuminates the world. Not a bad time to consider the ways in which the light of reason illuminates reality. Science illuminates nature but does not deplete its mystery. Science at its best—as practiced by a Galileo, a Herschel, an Einstein, or a Hubble—is an almost religious activity; a deliberate effort to engage intellectually, passionately with the mystery that permeates every particle of existence, every glimmer of light in the night sky. It was the encounter with mystery at the shore of knowledge that inspired Einstein's life work and reinforced his sense of the worthiness of human life. "Measured objectively," he wrote, "what a man can wrest from Truth by passionate striving is utterly Infinitesimal. But the striving frees us from the bonds of self and makes us comrades of those who are the best and the greatest." Einstein was proud of his Jewishness but open to the purest lights of every faith. The following letter he wrote to a group of children is self-explanatory.

Dear Children,

It gives me great pleasure to picture you children joined together in joyful festivities in the radiance of Christmas lights. Think also of the teachings of Him whose birth you celebrate by these festivities. Those teachings are so simple—and yet in almost 2000 years they have failed to prevail among men. Learn to be happy through the happiness and joy of your fellows, and not through the dreary conflict of man against man! If you can find room within yourselves for this natural feeling, your every burden in life will be light, or at least bearable, and you will find your way in patience and without fear, and will spread joy everywhere.

Is that enough? In a universe that is (seemingly) destined to infinite dispersal and darkness (in tens of billions of years), is it enough to "be happy through the happiness and joy of your fellows"? Einstein's theory of gravity—general relativity—and the theory of quantum mechanics, to which he contributed, add a strange twist to the universe's final days, or rather final gazillian years.

As time passes, the galaxies will go dark, and clusters of galaxies will drift ever farther apart, but within the clusters there will be continued consolidation, until more and more extinguished stars and galaxies coalesce into supergigantic black holes, knots of matter so dense that nothing can get out. But—and here's the twist—given enough time, it turns out that the laws of quantum mechanics allow black holes to evaporate. Energy and time are related in quantum mechanics by the so-called uncertainty principle, which describes a kind of fuzziness in nature, enough fuzziness to allow black holes to leak out particles over a long enough time. The times required for the evaporation of the gigantic black holes that will characterize the dying universe are unimaginably large—a number of years equal to 1 followed by one hundred 0s—hardly worth thinking about. In theory, at least, as black holes evaporate, the temperature goes up and the process of evaporation increases, eventually reaching runaway proportions. The thin, dark universe of the far, far future might be in for some fireworks!

So what? What has that got to do with the here and now? In Anton Chekhov's play *The Three Sisters,* sister Masha refuses "to live and not know why the cranes fly, why children are born, why the stars are in the sky. Either you know and you're alive or it's all nonsense, all dust in the wind." Why? Why? The striving to know is what frees us from the bonds of self, said Einstein. It doesn't really matter what is the fate of the physical universe or even the ultimate fate of life and intelligence. It's the striving to know, rather than our knowledge—which is always tentative and partial—that is important. Instead of putting computers in our elementary school classrooms, we should take the children out into nature, away from those virtual worlds in which they spend unconscionable hours, and let them see an eclipsed Moon rising in the east, a pink pearl. Let them stand in a morning dawn and watch a slip of comet fling its tail around the Sun. Let them admire the stars of Orion on a sparkling winter evening—red Betelgeuse, blue Rigel—and shiver in the thrall of cold and beauty.

"Either you know and you're alive or it's all nonsense, all dust in the wind," says Masha. Let the children know. Let them know that nothing, nothing they will find in the virtual worlds of E-games, television, or the Internet matters half so much as a glitter of stars on an inky sky, drawing our attention into the incomprehensible mystery of why the universe is here at all, and why we are here to observe it. The winter Milky Way rises in the east, one trillion individually invisible points of light, one trillion revelations of the Ultimate Mystery, conferring on the watcher a dignity, a blessedness, that confounds the dull humdrum of the commonplace and opens a window to infinity.

Appendix 1

Planets

Five planets are visible to the naked eye: Venus, Jupiter, Saturn, Mars, and Mercury. Venus and Mercury are always near the Sun, in the evening or morning sky. Venus is called "the evening star" or "the morning star" depending on its location. Mercury is difficult to see in the twilight or dawn; it is not described in the summaries that follow, but your newspaper and the "Resources" section will help you find it. Generally, the outer planets move eastward through the constellations of the zodiac, although occasionally they move "backward" in a motion called *retrograde*. This *apparent* reversal is actually due to the Earth's overtaking motion.

2001. Jupiter and Saturn are in Taurus as the year begins. Venus is an evening star early in the year, shining in the southwest. Mars rides with Scorpius in summer, near red Antares, then moves eastward to maintain its evening place in the southwest. Jupiter and Saturn return with winter, in Gemini and Taurus respectively.

2002. Mars, Jupiter, and Saturn are in the evening sky as the year begins, though Mars sets early. In May, Venus, Mars, Jupiter, and Saturn gather low in the west at sunset. By midsummer, only Venus remains for a while as an evening star. In late fall the evening sky is planet free, until Saturn rises with Gemini. Jupiter is not far behind in Leo.

2003. Jupiter reverses toward Cancer, then moves back into Leo by summer. Saturn is near the horn of Taurus as the year begins. In late summer, Mars rises with Aquarius, then moves into Pisces to dominate the evening sky right into winter, when Saturn returns in Gemini. At year's end, Venus moves back to the evening sky.

2004. Saturn has moved into Gemini but still commands the winter sky. Mars is in the southwest; Venus is close to the horizon at sunset; Jupiter rises near midnight. As the season progresses, Venus become more prominent as it climbs toward Mars. Jupiter rises earlier each evening to blaze in the southeast. With spring, Venus and Mars race through Taurus toward Saturn, then Venus turns back into the sunset. Mars and Saturn quickly follow, leaving Jupiter alone in Leo.

2005. Saturn still creeps through Gemini to adorn the winter sky. Jupiter has moved to Virgo; it is in the evening sky by April. Venus and Saturn meet in the sunset in late June as Venus returns as an evening star. By fall, Mars rises near midnight.

2006. Saturn in Cancer is well up by midnight. Mars is high in the south in Aries, and Venus is gone. With spring, Jupiter rises near midnight in Libra; it will increasingly dominate the evening sky. As summer comes, Saturn and Mars dive together into the sunset, leaving Jupiter alone. The year ends with no evening planets.

2007. Saturn returns as a winter visitor in Leo. By Valentine's Day, Venus is back in the evening sky. With summer, Jupiter returns late near Scorpius. In late June, Venus and Saturn have a sunset conjunction in the west, then fall into the twilight. Jupiter creeps low across the southern horizon. The year ends again with no naked-eye planets, until Mars rises in Gemini.

2008. Mars in Taurus and Saturn in Leo are winter's evening planets. With spring, Mars races across Gemini and Cancer to join Saturn in early July. Jupiter rises near midnight on summer nights in Sagittarius, then earlier each evening. Venus is an autumn and early winter evening star, joining Jupiter in November.

2009. Venus blazes in the southwest at sunset; by March, it drops back toward the Sun. Saturn creeps through Leo, the only evening planet during the first half of the year. In late summer, Jupiter returns in Capricornus. At year's end, Mars rises late in Leo.

2010. Mars in Cancer is our winter planet. Saturn rises in Virgo later in the evening. Venus returns with spring as an evening star. By summer, Venus, Mars, and Saturn decorate the western sky at sunset. They gather close low in the west in August, then sink into the sunset. Jupiter returns to the evening sky in winter, in Pisces.

2011. Saturn is in Virgo in the spring, as Jupiter glides into the sunset. The giant planet returns to the evening sky near midnight in the fall. Venus returns as an evening star in winter.

2012. Bright Venus is in the southwest at sunset; Jupiter is in the south also at sunset, in Aries. They join for a brilliant sunset show in March; the crescent Moon drifts by on March 25–26. Mars is in Leo; Saturn in Virgo. First Jupiter then Venus falls into the sunset; they are gone by summer. Mars chases Saturn through Virgo, passing in August. As winter comes, Jupiter rises in Taurus.

2013. Jupiter dominates the evening sky in Taurus. Saturn rises near midnight in spring, in Libra. As summer comes, Jupiter sinks into the sunset, as Venus returns to the evening sky. Saturn is low in the south; it joins Venus and a crescent Moon for a sunset show in early September. At year's end, Jupiter is in Gemini.

2014. Again, Jupiter reigns supreme through winter. Mars returns with spring in Virgo, and Saturn follows in Libra. The two planets join in Libra in late August, low in the southwest. As winter comes, Mars hugs the western horizon at sunset, but otherwise no naked-eye planets are in the evening sky.

2015. Venus returns as an evening star with the new year and stays till summer; it joins Mars in Pisces in late February. Jupiter is in Cancer, then in Leo. Venus pulls alongside Jupiter in July. Saturn drifts toward Scorpius. As winter's chill arrives, the evening sky is planet free.

2016. Jupiter rises late in Leo; by spring, it is prominent. Mars and Saturn return with summer, near the claws of Scorpius. Venus returns to the evening sky in winter but stays close to the horizon.

2017. Venus and Mars are prominent in the southwest at sunset, joined by a crescent Moon at the end of January. They sink into the sunset by spring. Jupiter returns with Virgo to grace summer nights. Saturn moves toward Sagittarius. As winter comes, there are no naked-eye evening planets.

2018. Venus is the evening star by March. With summer, Jupiter is in Libra. By August, Jupiter, Saturn, and Mars are low in the southern sky.

2019. Mars is prominent in the southwest as the year begins. As Mars slips into sunset with summer, Jupiter rises in the east, followed soon by Saturn. Venus returns but stays close to the Sun. It joins Jupiter very low in the southwest in late November, then catches Saturn there in mid-December.

2020. Venus is a brilliant evening star right through spring. With summer, Jupiter and Saturn return together in Sagittarius and preside till winter. Mars is a winter planet in Pisces.

2021. Jupiter overtakes Saturn as the year begins but is very low in the southwest at sunset. Mars dominates winter evenings as it moves through Taurus. Venus is an evening star by summer, near Mars in the northwest; they are in conjunction in mid-July, though close to the sunset horizon. Jupiter and Saturn return in the fall, in Capricornus. Venus hangs low in the southwest as winter comes.

Appendix 2

Meteor Showers

You might see a few meteors on any night, but certain times of the year are better than others—because of the so-called showers. A shower takes its name from the constellation from which the meteors seem to radiate. Usually, after midnight is the best time to look for meteors. The peak dates of showers are given here, but nights to either side of the peak can be fruitful as well.

January 4. Quadrantids. The radiant lies in the constellation Bootes, in a part of the sky once occupied by the eighteenth- and nineteenth-century constellation Quadrans Muralis, the Mural Wall.

April 21. Lyrids. A shower of variable reliability that can occasionally be good.

August 12. Perseids. The most consistently enjoyable shower of the year, with warm summer nights adding to the pleasure. You may see dozens per hour.

November 17. Leonids. Every thirty-three years or so the Leonids put on quite a show, as many as thousands of meteors per hour. Peak years were 1998 and 1999.

November 21. Alpha Monocerotids. The radiant lies near Procyon in Canis Minor. A variable shower that can be very strong.

December 13. Geminids. An excellent, reliable shower. You may see dozens per hour.

Glossary

asteroid One of several thousand very small members of the **solar system** that revolve around the Sun, generally between the orbits of Mars and Jupiter.

autumnal equinox The point on the **celestial equator** where the Sun crosses the equator moving south; also the time when the Sun is there (about September 21).

big bang The titanic explosion from a seed of infinitely dense energy out of which the universe is supposed to have had its beginning.

binary star A double-star system — two stars revolving about a common center of mass.

black hole A mass of such density that not even light can escape the pull of its gravity. Black holes are presumed to form from collapsed massive stars at the end of their lives. Black holes with masses of millions of stars are also believed to exist at the centers of many galaxies.

celestial equator Points on the **celestial sphere** directly above the Earth's equator.

celestial poles The two points on the **celestial sphere** directly above the poles of the Earth, marked approximately in the northern sky by the star Polaris; the intersection of the Earth's axis with the celestial sphere.

celestial sphere The imaginary sphere of the sky on which all celestial objects apparently reside; in former times, believed to literally exist.

constellation An arbitrary grouping of visible stars, taken to represent some familiar figure. There are eighty-eight official modern constellations.

dark matter A yet-unidentified substance or collection of objects, detected by its gravitational influence on visible objects, that may constitute the greater part of the mass of the universe.

ecliptic The Sun's apparent annual path across the celestial sphere, through the constellations of the **zodiac.**

ecliptic plane The plane of the Earth's orbit about the Sun; approximately the plane of the entire **solar system.**

galaxy A system of millions to hundreds of billions of stars, sometimes containing large amounts of dust and gas. Some galaxies, including our own **Milky Way Galaxy**, have a flat spiral form with a central bulge.

globular cluster A spherical distribution of tens of thousands of stars gravitationally bound to a **galaxy**.

light-year The distance light travels in a year at 186,000 miles per second — about 6 trillion miles.

Messier object An object listed in a catalog of nebulous celestial objects by Charles Messier in 1787.

meteor A track of light in the sky resulting from the vaporization of solid matter entering the Earth's atmosphere at high speed; commonly called "shooting stars" or "falling stars."

meteor shower Many meteors seen within a short period of time. Some showers occur at the same time each year as the Earth in its orbit passes through dusty regions of space.

Milky Way A band of faint light circling the **celestial sphere** that has as its source the myriad stars of the **Milky Way Galaxy**.

Milky Way Galaxy A **galaxy** of approximately one trillion stars that contains the Sun; it takes its name from the band of light in the sky called the **Milky Way** by the ancients.

nebula Any cloud of interstellar gas or dust.

neutron star A collapsed, extremely dense star consisting almost entirely of neutrons; the final state of a star about three to eight times as massive as the Sun.

nova A star that suddenly flares in brightness by a factor of hundreds of thousands.

occultation The eclipse of one celestial object by another, as when the Moon passes in front of a star or planet.

open cluster (or galactic cluster) A loose cluster of stars of common age and origin, found in the arms of a spiral galaxy. The Pleiades are the most familiar open cluster.

parallax The apparent change in the position of an object when viewed from two different locations; used to measure the distance to nearby stars.

precession The slow wobble of the Earth's axis as it spins, which causes the positions of the celestial poles and equator to change relative to the stars; one cycle of precession takes 26,000 years.

proper motion Real, not apparent, motion of a star across the sky relative to the positions of distant stars and galaxies.

quark One of the subatomic particles of which physicists suppose elementary particles such as protons, neutrons, and electrons are made. They existed independently only in the first instants of the **big bang.**

quasar (quasi-stellar radio source) An intense starlike source of radio energy and light, billions of **light-years** away and therefore seen early in the universe's history; the nature of quasars is poorly understood but may be related to the formation of massive **black holes** at the centers of galaxies.

radiant (of a meteor shower) The point in the sky among the stars from which meteors in a shower seem to radiate.

red dwarf A small cool star, less massive than the Sun.

red giant A cool red star, very bright because of its large size; a late stage in the life of a typical star.

solar system The Sun, planets, comets, and all other objects bound to the Sun by gravity.

spring equinox See **vernal equinox**.

summer solstice The point on the **celestial sphere** where the Sun reaches its northernmost position; also the time when the Sun is there (about June 21).

supernova An unusually violent explosion of a star, which results in an increase in brightness of hundreds of millions of times, or more; a typical fate of stars much more massive than the Sun.

tropic of Cancer Parallel of latitude on Earth, twenty-three and a half degrees north of the equator, corresponding to the Sun's northernmost position on the **celestial sphere.**

tropic of Capricorn Parallel of latitude on Earth, twenty-three and a half degrees south of the equator, corresponding to the Sun's southernmost position on the **celestial sphere.**

vernal equinox The point on the **celestial equator** where the Sun crosses the equator moving north; also the time when the Sun is there (about March 21).

white dwarf A hot white star that is small and dense; the final stage of stars with masses similar to the Sun's mass or smaller.

winter solstice The point on the **celestial sphere** where the Sun reaches its southernmost position; also the time when the Sun is there (about December 21).

zenith The point in the sky directly over an observer's head.

zodiac The twelve **constellations** that the Sun passes through during its annual apparent journey around the **celestial sphere.**

Resources

Books

Ottewell, Guy. *Astronomical Calendar.* This wonderful large-format book, published annually, is the sky watcher's invaluable companion. It will tell you everything that is going to happen in the sky worth looking for: the positions of planets, eclipses, meteor showers, and much more. Order from Universal Workshop, Furman University, Greenville, SC 29613.

Raymo, Chet. *The Soul of the Night: An Astronomical Pilgrimage*. St. Paul, Minn.: Hungry Mind Press, 1996. Meditations on the night.

———.*365 Starry Nights: An Introduction to Astronomy for Every Night of the Year*. New York: Simon & Schuster, 1982. More to see and more to imagine.

Schaaf, Fred. *Seeing the Sky: 100 Projects, Activities, and Explorations in Astronomy.* New York: John Wiley, 1990. A gold mine of information about everything from the brightness of twilight to the mysterious midnight gegenschein.

Software

Starry Night Backyard, for Macintosh and Windows, by SPACE.com Canada, Inc., 284 Richmond Street East, Suite 300, Toronto, Ontario, M5A 1P4, Canada, or www.SPACE.com. With a few clicks, you can see the sky on any night past, present, or future, from any place on Earth or elsewhere. A wonderful, user-friendly program of exceptional power. *Starry Night Pro,* an even more powerful version, is also available. I used it to make my planet projections and as a guide for preparing the star maps of this book. Highly recommended.

WEB SITES

oposite.stsci.edu/pubinfo/pictures.html Astonishing photos from the Hubble Space Telescope.

www.kalmback.com/astro/astronomy.html The Web site of *Astronomy* magazine.

www.skypub.com The web site of *Sky and Telescope* magazine. Lots of useful resources for following what's going on in the sky.

www.SPACE.com The definitive space Web site. From here, you can get to anywhere. Check out the list of space-related Web sites.

Information About the Photographs

PHOTOGRAPH ON PAGE 7

A deepest-ever view of the universe, showing hundreds of galaxies near the beginning of time, billions of light-years away. The image shows an area of the sky about the size of the intersection of crossed sewing pins held at arm's length. A deep view in any direction of the sky would show approximately the same thing.

Credit: R. Williams and NASA

PHOTOGRAPH ON PAGE 23

The Pleiades is a cluster of stars about 400 light-years away in our own neighborhood of the Milky Way Galaxy. The cluster is young, just tens of millions of years old, and still shrouded in wisps of the nebula out of which it was born. Typically, six of these stars can be seen with the unaided eye. Many of the faint stars in the photo are in the background.

Credit: David Malin and Anglo-Australian Observatory

PHOTOGRAPH ON PAGE 41

The central and outer region of a globular cluster, a jewel box of brilliant stars, thousands of them, as old as the Galaxy itself. Several dozen globular clusters buzz about the Milky Way Galaxy like bees. The curious shape of the photograph reflects the geometry of the Hubble Space Telescope's cameras. Four cameras image the cluster. One of them—imaging the cluster's core—has a smaller field of view but greater resolution.

Credit: Rebecca Elson, Richard Sword, and NASA

PHOTOGRAPH ON PAGE 59

The aftermath of a magnetic storm on the Sun, imaged by the Transition Region and Coronal Explorer (TRACE) satellite on November 8, 2000. Cooling material falls back onto the surface along lines of magnetic force. Flare-ups such as this cause aurora displays and radio disruptions on Earth.

Credit: Stanford-Lockheed Institute for Space Research and NASA

PHOTOGRAPH ON PAGE 77

Jupiter's volcanic moon Io passes above the turbulent clouds of the giant planet on July 24, 1996. Note the shadow of the moon on Jupiter's cloud tops.

Credit: J. Spenser and NASA

PHOTOGRAPH ON PAGE 95

The twenty-one icy fragments of Jupiter-fragmented Comet Shoemaker-Levy 9 strung out across 710,000 miles of space—three times the distance between the Earth and Moon—on course for a mid-July 1994 collision with the giant planet.

Credit: Hal Weaver and T. Ed Smith, and NASA

PHOTOGRAPH ON PAGE 115

Astronaut Edwin Aldrin walks on the Moon near a leg of the Lunar Module during the Apollo 11 mission. Neil Armstrong took the photograph. The astronauts' footprints are visible in the lunar dust.

Credit: NASA

PHOTOGRAPH ON PAGE 133

The huge star Eta Carinae (one hundred times more massive than our Sun) blows off lobes of gas as it approaches the end of its life. Each lobe is vastly larger than our solar system. The gas races outward at more than a million miles per hour. The star may finally die in a cataclysmic explo-

sion that expels all but the compact core into space.
Credit: Jon Morse and NASA

PHOTOGRAPH ON PAGE 151

Radiation from hot nearby stars evaporates gas from the giant Eagle Nebula in the Milky Way Galaxy. The globules that resist evaporation may harbor embryonic stars that hold the gas to themselves by gravity. Stars in the Galaxy are born from such nebulae, and give their matter back to the nebulae, enriched in heavy elements, when they die explosively.
Credit: J. Hester, P. Scowen, and NASA

PHOTOGRAPH ON PAGE 169

The majestic spiral galaxy NGC 4414, located about 60 million light-years from our Milky Way Galaxy. The galaxy was imaged by the Hubble Space Telescope as part of a project to refine the galactic distance scale.
Credit: NASA and the Hubble Heritage Team

PHOTOGRAPH ON PAGE 185

Astronauts Story Musgrave and Jeffrey Hoffman service the Hubble Space Telescope during the 1993 mission to repair the telescope's faulty vision.
Credit: NASA

PHOTOGRAPH ON PAGE 203

Two spiral galaxies collide. Trapped by their mutual gravity, they will continue to distort and disrupt each other. Eventually, billions of years from now, they will merge into a single more massive galaxy.
Credit: NASA and the Hubble Heritage Team

Index

OTHER BOOKS BY CHET RAYMO

NONFICTION

365 Starry Nights

The Crust of Our Earth

Biography of a Planet

The Soul of the Night

Honey from Stone

Written in Stone (with Maureen E. Raymo)

The Virgin and the Mousetrap

Skeptics and True Believers

Natural Prayers

FICTION

In the Falcon's Claw

The Dork of Cork

OTHER BOOKS BY CHET RAYMO

NONFICTION

365 Starry Nights

The Crust of Our Earth

Biography of a Planet

The Soul of the Night

Honey from Stone

Written in Stone (with Maureen E. Raymo)

The Virgin and the Mousetrap

Skeptics and True Believers

Natural Prayers

FICTION

In the Falcon's Claw

The Dork of Cork